To Mom and Dad,

with all my love and appreciation

ACKNOWLEDGEMENTS

The only words that I think aptly describe the last 10 years of my life are from Jerry Garcia, "What a long, strange trip it's been." I can only hope that there is life after education. It seems like it was just yesterday that I was leaving home for the first time to go to a strange place called college. I cannot even begin to comprehend all of the things that have happened to me since that day. But there are many people who I can thank for all their help, guidance, understanding, and friendship.

First and foremost, I would like to thank my research director, Dr. Eric Enholm, for all of his patience, guidance, and assistance. His help in the preparation of this manuscript was immeasurable. I would also like to thank my undergraduate research director, Dr. Gary Crowther, for introducing me to the wonderful world of chemistry. I would like to thank all of the professors in the division of organic chemistry who showed me that there were many facets to being a Ph.D. chemist. I would also like to thank Dr. Girija Prasad for training me and for his contributions to the unsaturated ketone project.

The thing that I will remember best and miss most about the University of Florida is the many friendships that I have developed since I have been here. Thanks go to Keith Palmer and Brent Kleintop for putting up with me this long and for showing me the mechanistic side of organic chemistry (Keith) and that analytic chemists are not just mechanics (Brent). Thanks also go to fellow graduate students and fantasy football friends Don Eades and Larry Villanueva for many drunken hours of Gator sports prognosticating. Additional thanks are extended to all of the past and present members of the Enholm research group, especially Janet Burroff, Hikmet Satici, and Jeff Schreier. I wish them all the best of luck in their careers and personal lives.

I am also grateful to Karin Larsen for making this last year very pleasant, despite the problems encountered at work. I wish that we had met earlier, but I hope we will continue to build on what we have begun. My family truly deserves the most appreciation though. Thanks go to my oldest brother, Mike, who showed that a Kinter could get his Ph.D., and to Chris for being an organic comrade in the family. Thanks go also to Jim and Tom for showing me that there is more to life than chemistry. But most of all, I am truly indebted to my parents. Without the emotional and financial support of my parents I would never have been able to achieve this goal. I feel very fortunate to have had these opportunities, which would not have been possible without the support of all of those persons mentioned above. For this I will always be indebted to them.

TABLE OF CONTENTS

Abstract of Dissertation Presented to the Graduate School
of the University of Florida in Partial Fulfillment of the
Requirements for the Degree of Doctor of Philosophy

CARBONYLS AS FREE-RADICAL PRECURSORS:
CYCLIZATIONS OF UNSATURATED AND EPOXY CARBONYLS

By

Kevin S. Kinter

May, 1993

Chairman: Eric J. Enholm
Major Department: Chemistry

This dissertation investigated the tributyltin hydride
induced cyclization reactions of unsaturated ketones and
epoxy carbonyls with olefins. The reaction of tributyltin
hydride with aldehydes and ketones produces an O-stannyl
ketyl, which has both anionic and radical character. The
goal of this study was to ascertain whether the radical
reactivity of the O-stannyl ketyl could be delocalized away
from the central-carbon atom by the participation of labile
functional groups.

The first area of study was the intramolecular coupling
of unsaturated ketones with alkenes to produce functionalized
cyclopentanes. Unactivated alkenes were found to be
unsuitable as radical acceptors and activation of the alkene
was essential to the cyclization. A dilution study revealed

that excellent anti stereoselectivities (>50:1) could be achieved, and this was attributed to a reversible cyclization. Product identification and enolate trapping studies demonstrated that the anionic character of the tin ketyl could also be utilized. This was the first reagent-based approach to study the coupling of the β-carbons of these systems.

The second area of study investigated the cyclizations of epoxy carbonyls with olefins. The C-O bond of the epoxide was fragmented by the adjacent tin ketyl radical to produce an alkoxy radical. This project was originally designed to produce tetrahydrofurans, however these reactions produced functionalized cyclopentanols. Originally it was believed that the reactions were proceeding through a hydrogen abstraction mechanism, but after further experimentation a rare 1,5-tin transfer mechanism was proposed.

CHAPTER 1

INTRODUCTION

The term "free radical" applies to a species which possesses one unpaired electron.[1] Free radicals were once thought to be indiscriminate, highly reactive intermediates, but lately are being viewed in a kinder, gentler light.[2] Radicals are formed by a homolytic cleavage of a covalent bond. The central atom is usually sp^2 hybridized and the unpaired electron rests in the p-orbital.[3] The reactivity which radicals exhibit is dominated firstly by the nature of the central atom and secondly by the substituents which are attached to it.[2]

Free radicals have been known since 1900 when Gomberg investigated the formation and reactions of triphenylmethyl radical.[4,5] Paneth not only discovered that less stable alkyl radicals exist, but also measured their lifetimes in the gas phase.[6] Synthetic applications of radicals began in 1937 when Hey and Waters described the phenylation of aromatic compounds by dibenzoyl peroxide.[7] In the same year Kharasch proposed that the anti-Markovnikov addition of hydrogen bromide to an alkene was also a radical chain process.[8] Until the 1970s, synthetic free radical chemistry developed

1

slowly and was mostly applied only to copolymerization reactions.[9,10]

In the 1970s though, new synthetic methods involving free radicals began to be developed,[11,12] and lately radical cyclizations have become integral parts of many elegant syntheses.[13] The power and versatility of radical chemistry is best demonstrated by Dennis Curran's synthetic work towards the class of natural products known as triquinanes.[14] Curran and Rakiewicz's landmark synthesis of hirsutene 2, as shown in Scheme 1-1,[15] utilizes a tandem radical cyclization approach as the key step. The linear triquinane was produced in a single step by a tandem sequence commencing with the

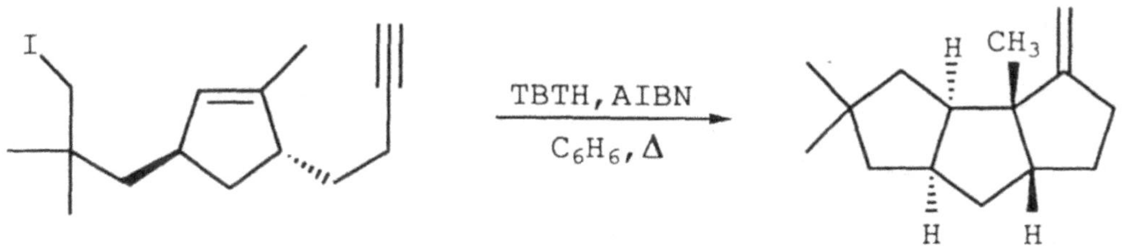

Scheme 1-1

generation of a 5-hexenyl radical from primary iodide 1 which was captured by the olefin and finally terminated by addition to the suitably disposed alkyne. Curran has also used tandem radical cyclizations in the synthesis of other members in the triquinane family (Figure 1-1), such as capnellene,[16] coriolin,[17] modhephene 3,[18] silphiperfolene 4,[19] and hypnophilin 5.[17]

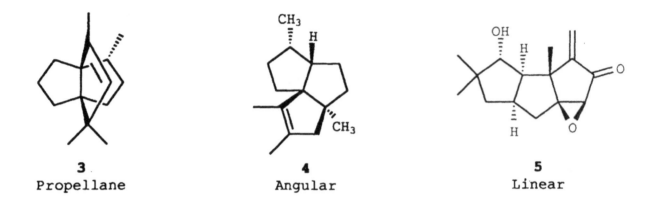

Figure 1-1

Other triquinanes synthesized by Curran

Before one can appreciate the complexity and grace of the aforementioned syntheses one must understand the basic principles of these reactions. Radical processes typically are chain reactions, and the first element of a chain reaction is initiation. Many reagents which will begin a radical chain reaction are available to the synthetic chemist.[2] The reagent of choice, from both a safety and a practical standpoint, is azobisisobutyronitrile (AIBN) 6, which is thermally decomposed to give cyanoisopropyl radicals (Scheme 1-2). These radicals are not reactive enough to abstract alkyl's hydrogens, but they are capable of abstracting a hydrogen atom from the weak Sn-H bond of tributyltin hydride (TBTH) 9 to produce the very useful tributyltin radical 11.[2]

For more than 20 years tributyltin hydride has been known to engage in free radical reactions.[20] The chemistry that TBTH engages in is too extensive to cover here, but good reviews of its reactions are available.[1,13,21,22,23] It is commercially available and it can be prepared from bistributyltin oxide and polymethylhydoxy siloxane.[24] Lately, there have also been some tin hydride reagents developed which are immobilized on polystyrene beads.[25] Although there are many different initiators and reagents, this discussion will focus mainly on the AIBN/TBTH system which was employed in my work. The majority of current radical reactions are based on the chemistry of this reagent combination.[22]

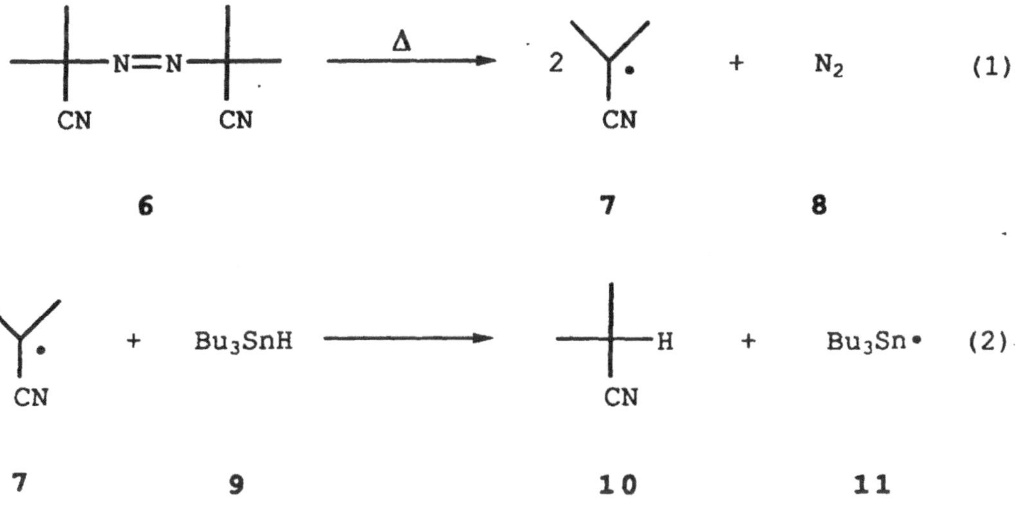

$$\text{6} \xrightarrow{\Delta} 2 \text{ 7} + N_2 \quad (1)$$

6   **7**   **8**

$$\text{7} + Bu_3SnH \longrightarrow \text{10}-H + Bu_3Sn\cdot \quad (2)$$

7   **9**   **10**   **11**

Scheme 1-2

For a chain reaction to develop, the radicals formed in the initiation steps must propagate. Propagation (Scheme 1-3) occurs when a radical 11 interacts with a nonradical 12 to produce a new radical species 13. Without propagation the chain would never develop because if two radicals combine to form a nonradical, the process is terminated. Careful control of reaction variables, such as concentration and the substrate to initiator ratio, helps to avoid such nonproductive reactions. The organotin radical which is produced is not synthetically useful until it reacts with other functional groups to generate an organic-centered radical.

$$Bu_3Sn\bullet \quad + \quad R-X \quad \longrightarrow \quad R\bullet \quad + \quad Bu_3Sn-X \quad (3)$$

11   **12**   **13**   **14**

X= Halogen, -SR', -SeR', $-NO_2$

$$Bu_3Sn\bullet \quad + \quad \equiv \quad \longrightarrow \quad (4)$$

11   **15**   **16**

$$Bu_3Sn\bullet \quad + \quad X= \quad \longrightarrow \quad Bu_3Sn—X— \quad (5)$$

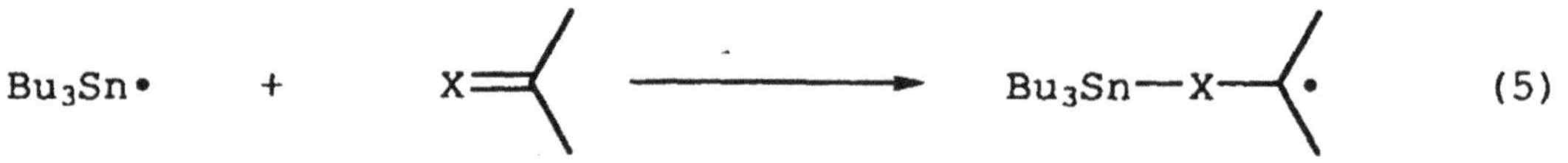

11   **17**   **18**

X= C, O, S

Scheme 1-3

There are two broad classes of free radical reactions: atom or group abstraction (eq. 3) and addition to multiple bonds (eq. 4 and 5). Halide abstraction was discovered in 1957,[26] and has achieved great importance in organic chemistry since. Atom abstraction involves an S_H2 reaction by tin radical to generate an organic radical. Bromine and iodine are most commonly used because chlorine reacts sluggishly and fluorine is unreactive. The degree of substitution of the carbon which bears the halide also has an effect on the rate of this reaction, which is as follows: $1^\circ RX < 2^\circ RX < 3^\circ RX$.[27] This trend parallels the relative stability of the resultant alkyl radicals.

Thiols and thioethers can also be reduced cleanly with TBTH, due to the strong tin-sulfur bond which is formed.[21] Thioketals can also be reduced in the same manner, which offers an interesting alternative for ketone reduction.[28] The most useful of the sulfur-related reductions involves the addition, fragmentation reactions of thiocarbonyls which were developed by Barton and McCombie.[29] This deoxygenation methodology was developed because of the inactivity of alcohols towards TBTH reductions. This method can be applied to a large variety of hydroxy compounds, including primary, secondary, tertiary, and diols.[21] It has been used in the synthesis of many natural products such as compactin,[30] anguidine,[31] and gibberellin.[32]

Another way in which tin can generate a carbon-centered radical is by its addition to multiple bonds. When 11 adds to alkenes or alkynes the ensuing radical 16 or 18 can then abstract a hydrogen from TBTH, giving the hydrostannated products 19 and 20. In the process another molecule of tributyltin radical, which satisfies the last criterion for a radical chain reaction, has been generated. When a terminal alkyne is reacted, the tributyltin will add to the terminus. The reaction proceeds through overall anti-addition, which gives the Z olefin as the kinetic product, but excess tin can equilibrate the reaction mixture so that the thermodynamically favored E-olefin is produced.[21]

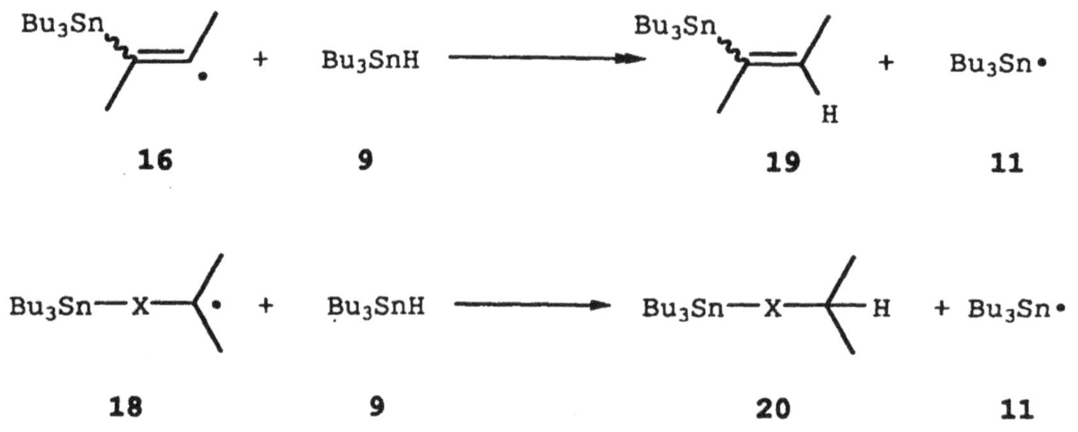

Scheme 1-4

The chemistry which is most important to this thesis is how tin reacts with carbonyls. Researchers have postulated that the hydrostannation of a ketone or aldehyde carbonyl can occur by two different mechanisms, depending on the reaction conditions, shown in Scheme 1-5.[21] When polar solvents and Lewis acid catalyst are utilized the ionic pathway dominates (eq. 6). In this mechanism, TBTH acts as a true "hydride donor" giving intermediate 22 which reorganizes to give the tin alkoxide. An alcohol can be obtained when 23 is treated either with proton sources or with additional TBTH.[21] Silica gel[33] and tributyltin triflate[34] have proved to be valuable catalysts.

An alternative mechanism is the way in which TBTH reacts with carbonyls through a free radical pathway (eq. 7). The formation of the O-stannyl ketyl 24 in this case arises from

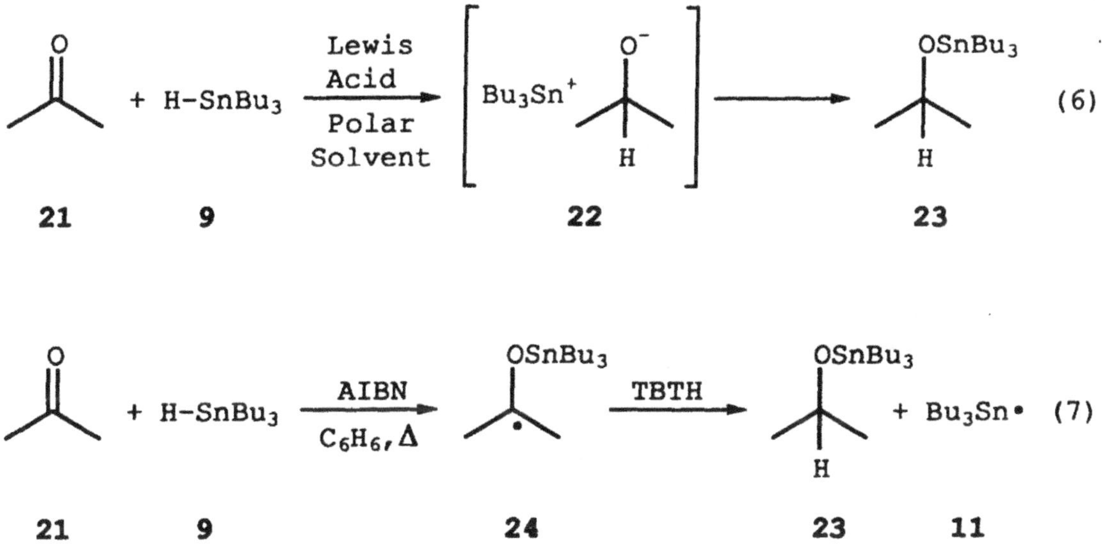

Scheme 1-5

the tributyltin radical reacting at the oxygen of the carbonyl to form a carbon-centered radical. This radical then abstracts a hydrogen atom to produce the tin alkoxide species 23, and the tributyltin radical can now repeat the process. It is this second type of reactivity, radical addition to carbonyls, which will be the principal focus of this dissertation.

Historically, rarely studied O-stannyl ketyls were first described by Tanner et al.,[35] and later by both Beckwith and Roberts[36] and Sugawara et al.[37] in the synthesis of multiple ring systems, discussed later. The tin ketyl can be considered a pseudo-protected radical anion, where the O-Sn bond has a large degree of ionic character due to electronegativity differences (Scheme 1-6). The apparent challenge to synthetic chemists is to develop methodology which exploits both elements of reactivity which are presented by this radical anion. Some of the work in Chapter 2 will address this challenge.

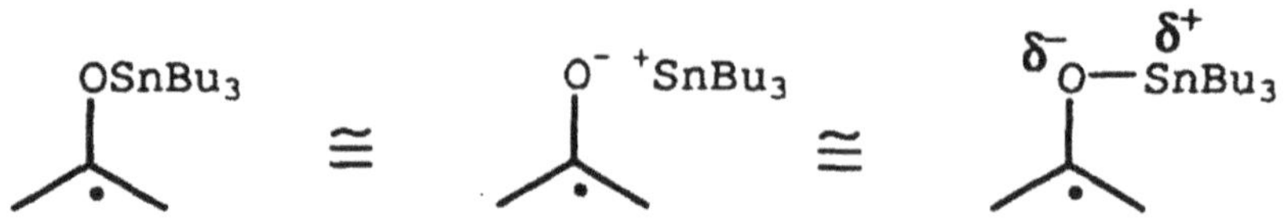

Scheme 1-6

So far, this dissertation has discussed reductive free-radical chain reactions and how they can produce carbon-centered radicals, followed by hydrogen atom abstraction. The next logical step is to examine their use in the formation of carbon-carbon bonds, which is particularly important because carbon-carbon bonds are the heart of organic synthesis. Before one can use radicals in the synthesis of complex molecules one must understand the physical organic principles of free radicals. Some of the factors which dictate how a radical will react include orbitals, conformations, thermochemistry, and reaction rates.

The basic premise upon which thermochemistry is built is that reactions which are exothermic (downhill in energy) are favorable processes. Very often the behavior of radicals can be rationalized from this standpoint. A good example is the well known ability of oxygen-centered radicals to abstract hydrogen. In this reaction a very strong O-H bond (~111 kcals)[3] is formed at the expense of a weaker C-H bond (~99 kcals).[3] Because a reaction is exothermic does not guarantee that it will happen spontaneously, but it can be used as a guide to see how feasible a transformation is.

The substituents on a radical also have a profound effect on a radical's behavior. Giese noted that cyclohexyl radical adds 8500 times faster to acrolein than to 1-hexene.[38] In contrast, trifluoromethyl radical undergoes addition reactions most efficiently with electron rich olefins such as enamines and enol ethers.[2] These results can be explained using frontier molecular orbital theory.[39] The singly occupied molecular orbital (SOMO) of a radical interacts with either the lowest unoccupied molecule orbital (LUMO) (Figure 1-2) or with the highest occupied molecule orbital (HOMO) (Figure 1-3) of the alkene.[2] In the trifluoromethyl example, the inductive, electron-withdrawing effect of the fluorines lowers the SOMO energy such that there is better orbital overlap with the low-lying HOMO of the alkene.

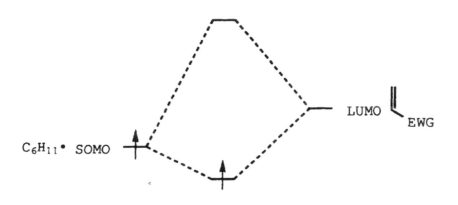

Figure 1-2

Reaction of cyclohexyl radical with electron-poor alkene

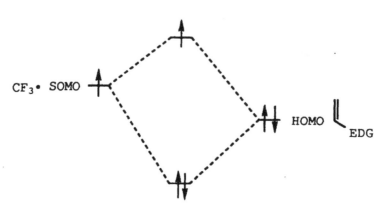

Figure 1-3

Reaction of trifluoromethyl radical with electron-rich alkene

Another factor which influences the regioselectivity and the stereochemistry of radical cyclizations is molecular geometry. The regioselectivity of radical transformations often complements or improves on the results which are available through ionic reactions. 5-hexenyl radicals (29) cyclize predominately to give five-membered rings (Scheme 1-7),[40] whereas cationic cyclizations yield six-membered rings.[1] Conjugate addition of radicals to activated olefins occurs solely at the ß-carbon, but anionic conjugate addition often shows a competition between 1,4 and 1,2 addition.[1]

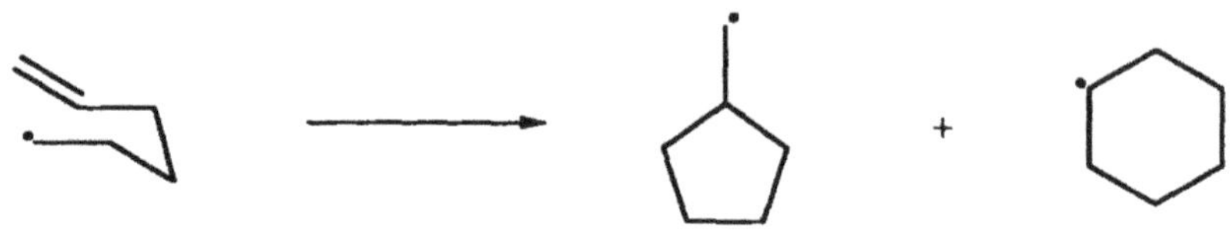

50   to   1

29   **30**   **31**

Scheme 1-7

12

The stereochemistry of 5-hexenyl radical cyclizations
has been thoroughly investigated by the Beckwith group.[41] In
many additions to double bonds, such as cationic, attack
occurs at the center of a double bond, but radical reactions
are generally accepted to proceed via an early, unsymmetrical
transition state.[42,43] Beckwith has formulated a model based
on the cyclohexane chair transition state (Figure 1-4) where
major products will be formed from the conformer with
substituents in the equatorial positions (Scheme 1-8).

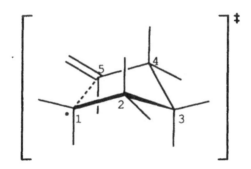

Figure 1-4

Beckwith's chair-like transition state

Once the factors which influence radical reactivity have
been considered, the reactions in which they participate can
be examined. These reactions can be divided into two main
categories: intermolecular and intramolecular. The
intermolecular reactions are more difficult to accomplish
because the radicals must find the other reactive partner
before they are reduced by TBTH. This problem can be
circumvented by using either syringe pump or hydrideless tin
techniques to maintain a low hydrogen donor concentration.[44]

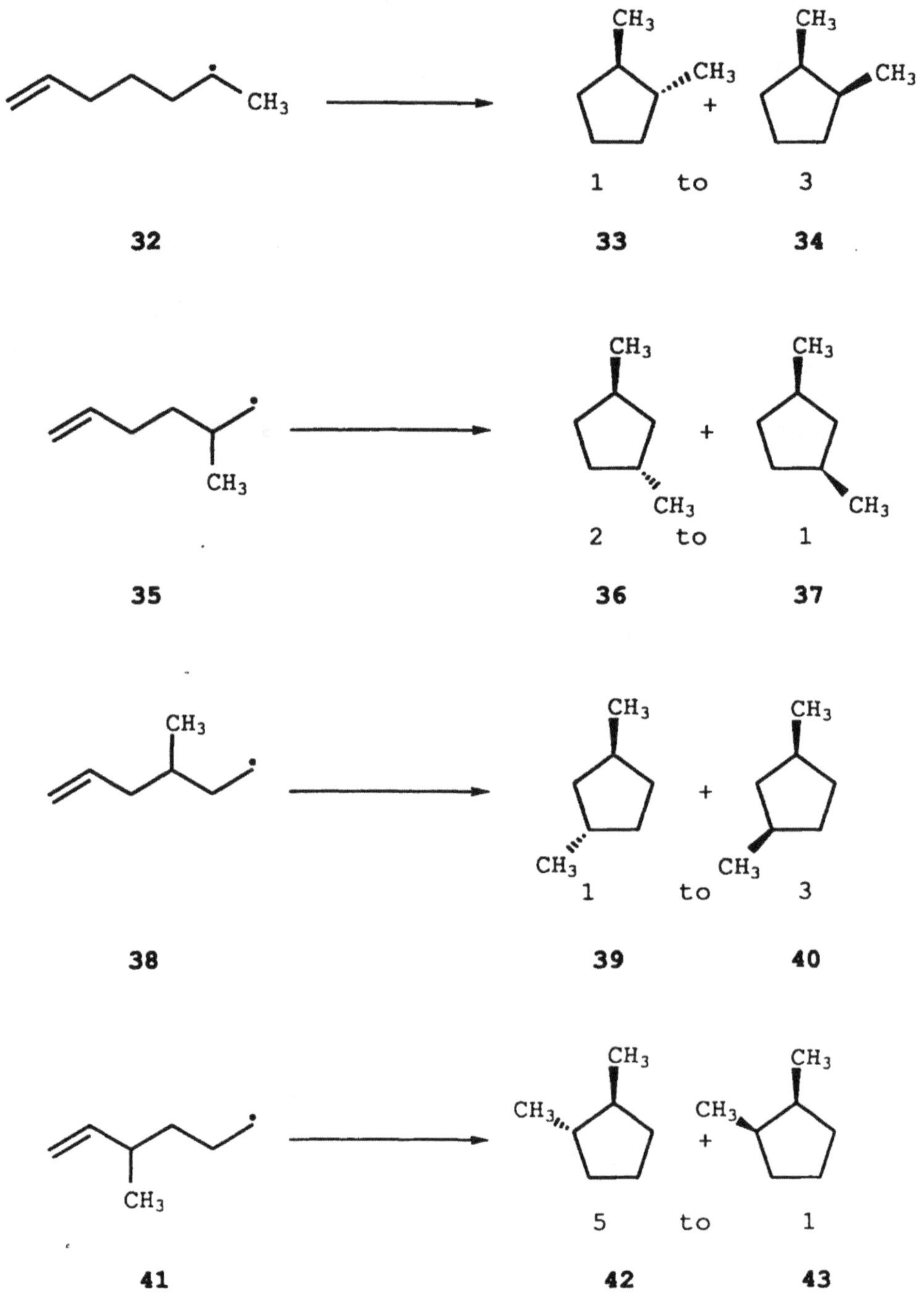

Scheme 1-8

The most basic intermolecular reactions that can occur are recombination and disproportionation, which are both radical-radical reactions. Recombination of radicals has found some applications in synthetic chemistry. The Kolbe electrolysis of carboxylate salts can lead to dimerization,[45] or mixed couplings if one acid is used in excess.[46] Disproportionation is the other pathway that two radicals can take when they meet. Disproportionation and recombination both occur when proponic acid is electrolyzed: butane is the product of recombination, and ethane and ethylene are the results of disproportionation.

Many intermolecular processes are simple substitution reactions where the intermediate radical is quenched with either a hydrogen source or a functional group. Dehalogenation and deoxygenation reactions have become a standard reaction in the armory of synthetic chemists[2] and have already been discussed in this dissertation. When the thiocarbonyl methodology is applied to carboxylic acids, the carboxylate can be replaced by many groups such as halogens, chalcogens, and phosphorus groups.[2]

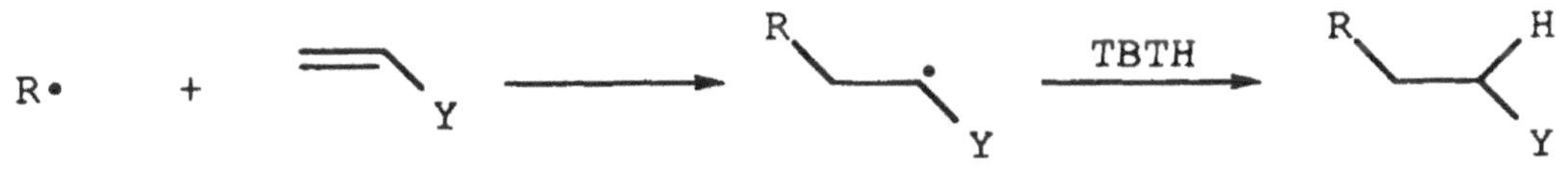

$$R\cdot \quad + \quad \diagup\!\!\diagdown_Y \quad \longrightarrow \quad \overset{R}{\diagdown}\!\!\diagup\!\!\cdot_Y \quad \xrightarrow{TBTH} \quad \overset{R}{\diagdown}\!\!\diagup\!\!\overset{H}{\diagdown}_Y$$

13	**44**	**45**	**46**

Scheme 1-9

The most important intermolecular methodology for the synthesis of aliphatic carbon-carbon bonds via radical reactions is the addition of alkyl radicals to alkenes[1] (Scheme 1-9). For these reactions to be successful there is a selectivity requirement which must be fulfilled. The selectivity requirement pertains to the intermediate radicals **11**, **13**, and **45**. These transient radicals all have specific partners that they must react with for a successful synthesis to occur. If **45** and **13** have the same tendency to add to alkenes, then the reaction will result in polymerization.[1]

This problem can be avoided by careful selection of the alkene appendage so that **13** and **45** have different selectivities. To avoid polymerization the electronic characteristics of the substituents on **13** and **45** must be opposite in nature. An electron donating group on **13**, (such

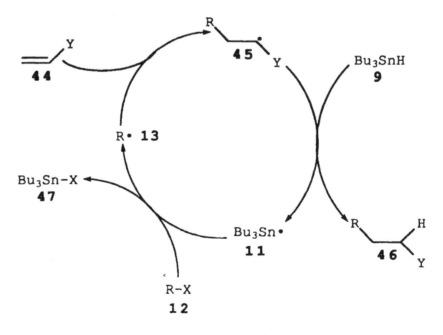

Scheme 1-10[1]

as a simple alkyl radical) and an electron withdrawing group on the alkene is a popular combination (eq. 8).[47] When these appendages are used 13 becomes a nucleophilic radical which prefers to add to electron deficient alkenes such as 49. This gives 45 (Y=CN) which is now an electrophilic radical because of the cyano group. Electrophilic radicals favor adding to electron rich alkenes, therefore it will not add to 49 (polymerization) and it will eventually be quenched by TBTH. Equation 9 shows the other potential combination of radical and olefin substituents.[48]

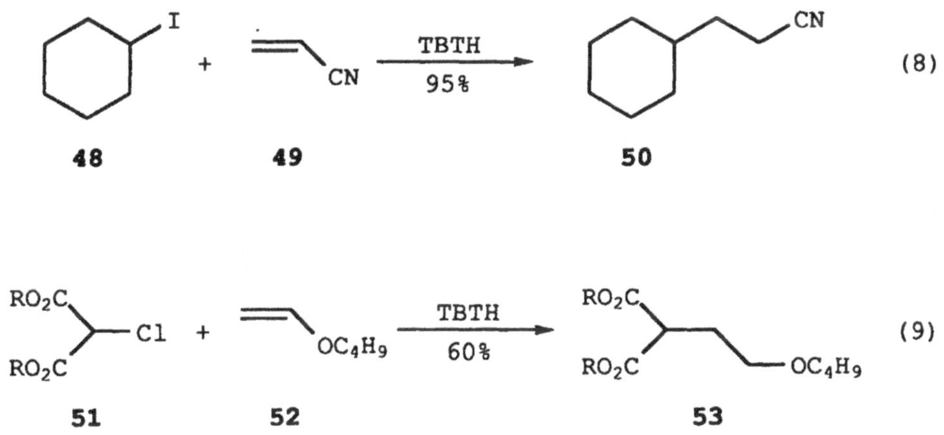

Scheme 1-11

One of the main criticisms regarding radical chemistry has been that due to the planarity of the intermediates their application in chiral synthesis is impossible. Motherwell rebuffs this argument by saying that "the planarity of enolate anions, iminium ions and even free carbocations has not hindered the effective operation of stereoelectronic

control elements leading to stereospecific reactions."[2]
Currently there are at least three groups working on the
control of acyclic stereochemistry, where the element of
control centers around the use of a chiral auxiliary on the
alkene. The Porter[49] and Giese[50] groups use the
dimethylpyrrolidine 54, and Curran and coworkers[51] use sultam
55 as their chiral auxiliary. Scheme 1-12 shows how Porter,
Scott, and McPhail used the dimethylpyrrolidine auxiliary to
control the addition of cyclohexyl radical to alkene 56 which
produced 58 as the only diastereomeric product.[52]

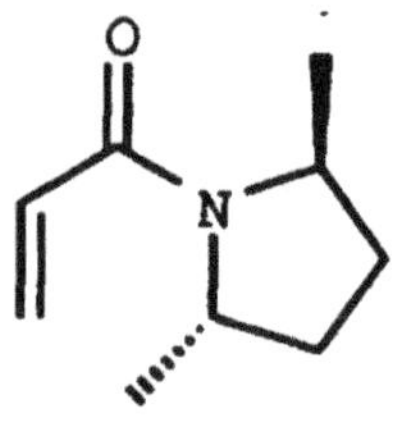

54 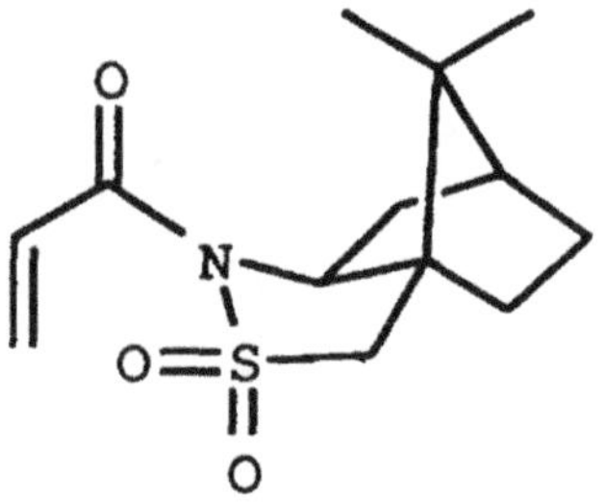   55

Figure 1-5

Chiral auxiliaries used by Porter, Giese, and Curran

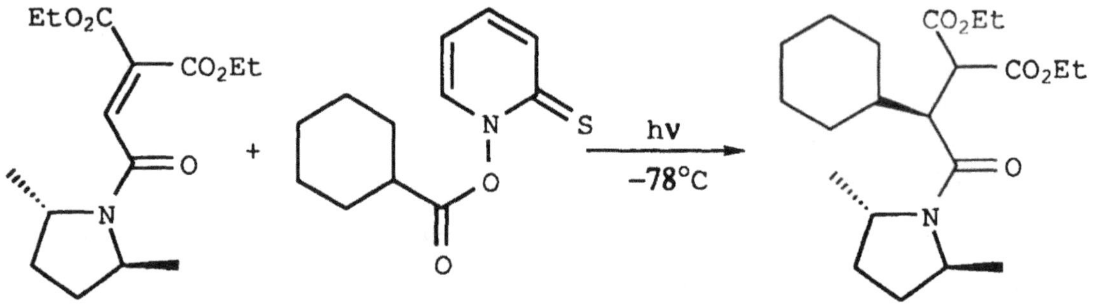

56   57   58

Scheme 1-12

The intramolecular reactions that radicals can undergo
can be grouped into two different categories: rearrangements
and cyclizations.   Rearrangements can take on many forms
ranging   from   simple   hydrogen   abstraction   to
cyclopropylcarbinyl rearrangements.   A popular intramolecular
reaction of radicals is a 1,5-hydrogen migration, which
proceeds through a six-membered transition state.[53] An
interesting example of this reaction is by Rawal and
coworkers[54] who recently reported the transformation shown in
Scheme 1-13. The olefin that is formed when the epoxide opens
eventually cyclizes with the carbon radical that is

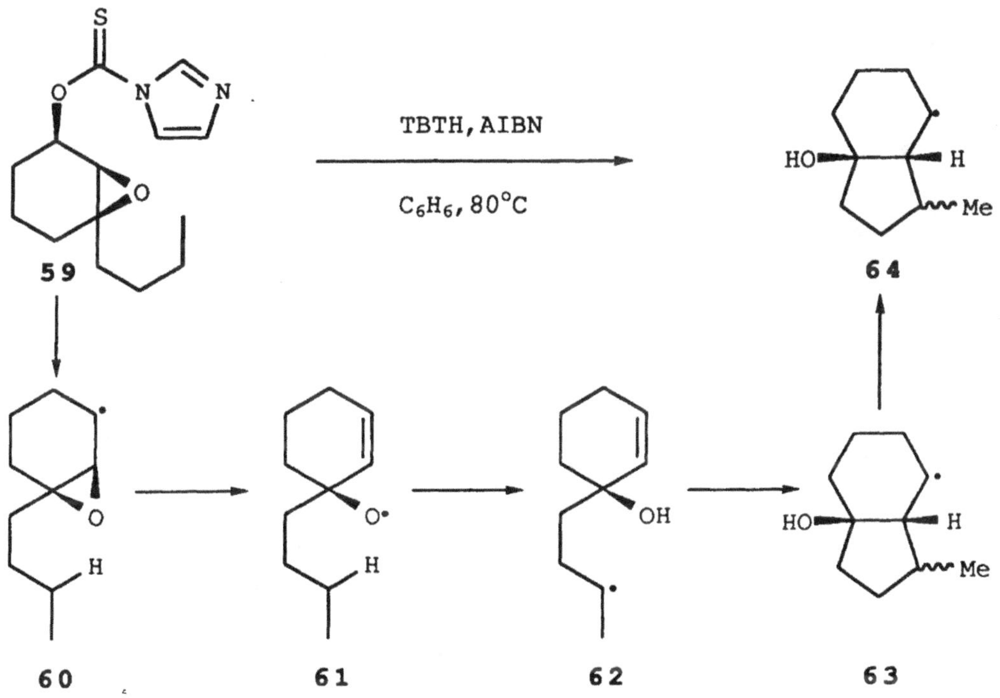

Scheme 1-13

formed when the alkoxy radical abstracts a hydrogen from the δ carbon. The analogous 1,5-tin migration will be the subject of Chapter 3 of this dissertation.

Carbon-carbon bonds can also be fragmented using radicals. When a radical is generated α to a cyclopropane it rapidly equilibrates to its homoallylic counterpart.[2] This ring-opening reaction has been used by Motherwell and Harling[55] to build bicyclospiro fused compounds (Scheme 1-14). The radical which is formed by the fragmentation of the thiocarboxylate opens the cyclopropane ring. This radical then cyclizes onto the suitably disposed alkyne, and the vinyl radical is eventually quenched by TBTH.[56]

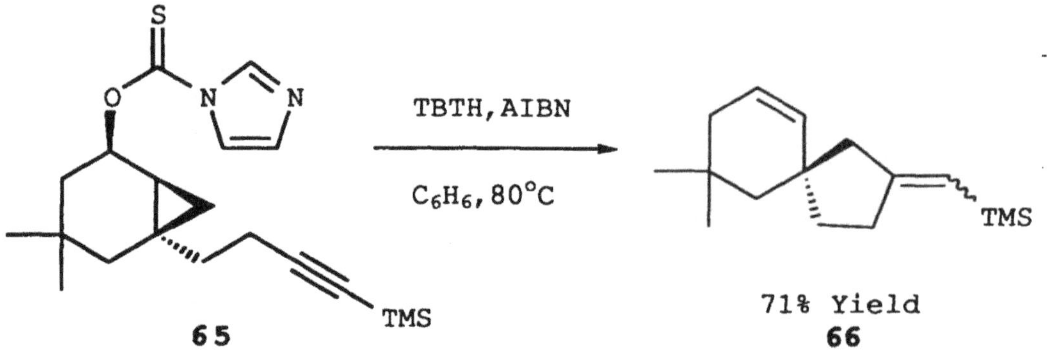

Scheme 1-14

As you can see from the examples above, radicals can abstract hydrogens and fragment C-C bonds, but their main function often lies in cyclization reactions. This thesis will examine how carbonyls can be used to generate carbon-centered radicals and the cyclizations that the resultant radicals undergo. Prior to Enholm's studies very few papers approached these reactions from a reagent-based standpoint. The photolysis[57] and electrochemistry[58] of ketones differ from the studies in this dissertation in that both are nonreagent based methods which will be only briefly reviewed here.

When carbonyls are irradiated with light, an $n-\pi^*$ transition occurs in which an electron from the nonbonding orbital on the oxygen is promoted to π-antibonding orbital of the ketone. The carbonyl in its excited state can either abstract a proton from the solvent or an available hydrogen donor, or it can fragment to give more stable molecules. Fragmentation reactions can be divided into two categories, type-I and type-II, based on the work of Ronald Norrish who shared the Nobel prize in chemistry in 1967. In type-I fragmentation reactions a carbonyl substituent is cleaved to produce acyl and alkyl radicals. In most cases the acyl radical is cleaved again to generate another alkyl radical and carbon monoxide. The photolysis of 67 by Quinkert and coworkers[59] demonstrated this reaction, where subsequent cleavage of the carbonyl appendages produced carbon monoxide and a diradical species which coupled with itself to form 68.

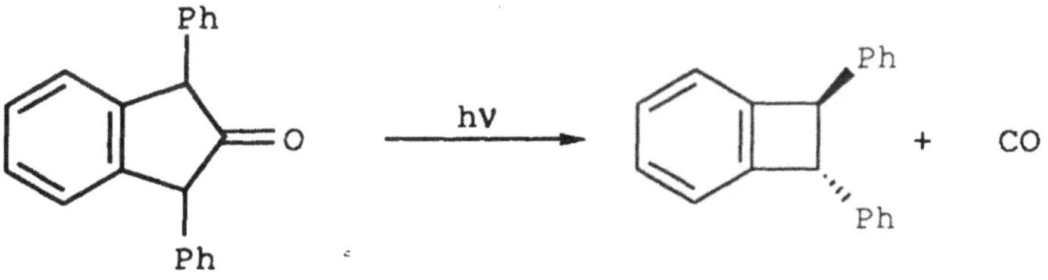

Scheme 1-15

The second type of Norrish cleavage happens when the
carbonyl compound being irradiated contains hydrogens on the
δ-carbon. In this reaction the photoexcited carbonyl
abstracts a hydrogen from the δ-carbon through a six-membered
transition state to give a carbon-centered radical. The
diradical which is produced, 72, can either fragment to give
74 and 75 or couple to yield cyclobutanols.[60] These
fragmentations can be distinguished by the bonds which are
broken. In type-I reactions the bond between the carbonyl and
the α-carbon is ruptured, but in type-II the C_α-C_β bond is
broken.

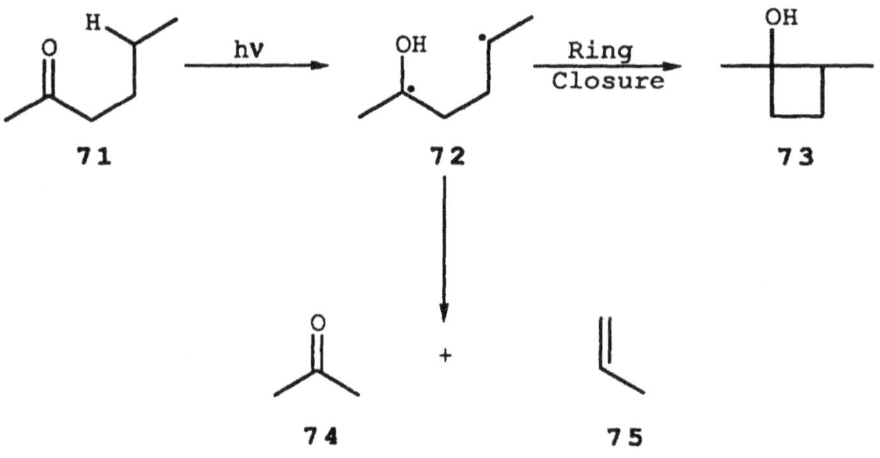

Scheme 1-16

The electrochemistry of carbonyls has also shown that
they can effectively act as radical precursors.[58] When
conducting solutions of ketones and aldehydes are
electrochemically reduced, there is a competition between
reduction to the alcohol and coupling of the intermediate

22

radical to give pinacol products. In recent years, however, there have been successful attempts to trap the ketyl radicals with a variety of appendages, such as allenes, alkenes, alkynes (Scheme 1-17),[61] and even benzene rings.[58] The nature of the reactive species is sometimes clouded by the conditions of the reaction, but it is generally believed that either a ketyl (radical anion) or a protonated ketyl is the reactive intermediate.

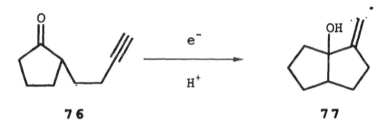

76   **77**

Scheme 1-17

Although sparse, there were two reports of the O-stannyl ketyls being used in free radical cyclizations. In the first, Beckwith and Roberts[36] showed that tin ketyls can be used to assemble bi- and tri-cyclic systems. The cyclization (Scheme 1-18) proceeded with excellent yield, but they remarked that the reaction was "sluggish" and required additional TBTH and AIBN. Beckwith speculated that the formation of the tin ketyl may be reversible or that either the rate of the radical's initial formation or its subsequent cyclization was slow. More likely, the problem with this cyclization is that the nucleophilic tin ketyl would prefer to react with a more electrophilic double bond.

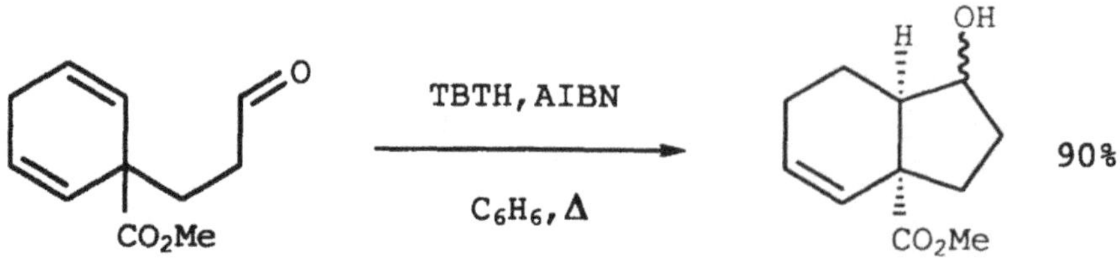

Scheme 1-18

More recent research indicates that cyclizations of tin ketyl are effective and have potential applicability in synthesis. In a second paper which preceded this work, Sugawara and coworkers[37] may have realized that an activated alkene would solve the problems that Beckwith observed. They used tin ketyls in the cyclization of aldehyde 80 onto the activated olefin of a uridine ring system (Scheme 1-19). Note that in this example, the skeleton is constrained to prepare a six-membered ring, and the 6-exo closure of the 6-heptenyl radical is known to be more than one order of magnitude slower than the analogous 5-hexenyl cyclization.[62] So an activated alkene is sometimes crucial to the success of a radical cyclization.

Aside from these two synthetic reports, only the Enholm group is actively reporting on studies directed towards the use of tin ketyls in synthesis. This interest began in 1989 when Enholm and Prasad[63] demonstrated that aldehydes and ketones readily cyclized onto tethered olefinic appendages. Although an unactivated olefin did cyclize, the yield was low, and to achieve a synthetically useful yield the olefin needed electron withdrawing groups, i.e., activation. They also demonstrated the 6-exo cyclization of an activated olefin, as shown in Scheme 1-20, where they produced cyclized products in a 69% yield.

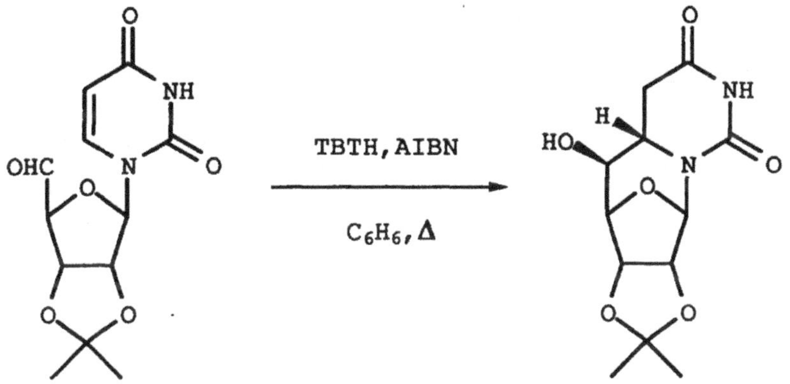

80   **81**

Scheme 1-19

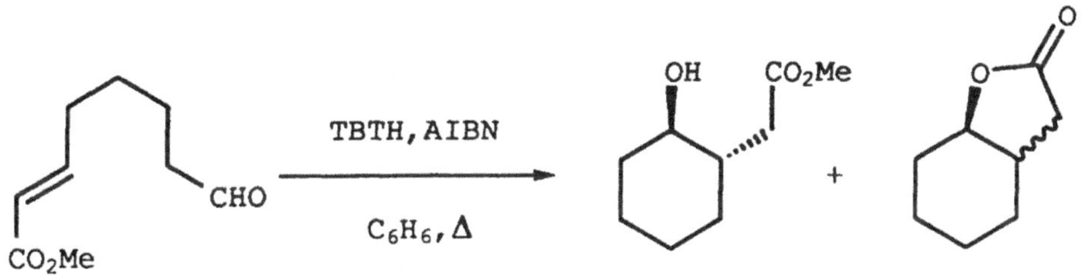

82   **83**   **84**

Scheme 1-20

The next step for the Enholm group was to see if tandem cyclizations were possible with tin ketyls. Enholm and Burroff[64] found that both spiro and fused bicyclic ring systems could be synthesized with this methodology. In this

study activated olefins were used in the first cyclization
and unactivated olefins were used in the second cyclization.
The activated olefin which is involved in the first
cyclization must act as both an acceptor and a donor in this
reaction. Also notice that intermediate radical **86** is
electrophilic and therefore needs a nucleophilic olefin for
effective cyclization. All of these criteria are met and the
reaction proceeds smoothly to give bicyclic products in a 75%
yield.

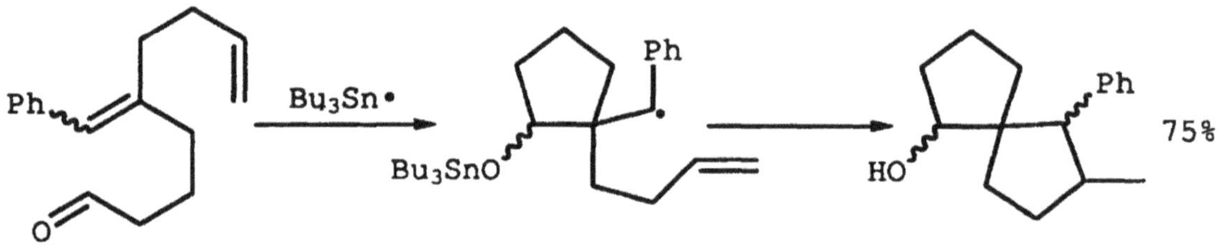

Scheme 1-21

This dissertation investigated the free radical behavior
of tin ketyls which are functionalized so that the radical
could be delocalized or separated from the carbonyl. This
was accomplished by the use of unsaturated and epoxy ketones.
TBTH was used to generate the tin ketyl, and activated and
unactivated olefins were used to trap the radical
intermediates.

In Chapter 2 the behavior of α,β-unsaturated ketones and
their cyclizations and additions to activated olefins was
examined. Highly functionalized monocyclic and bicyclic

cyclopentanes were obtained. An interesting concentration effect was observed which can be attributed to the reversibility of the cyclization. This concentration effect enabled the reactions to achieve excellent diastereoselectivities (>50:1). Both intramolecular and intermolecular additions were examined and their applicability towards five-membered ring natural products was considered.

Chapter 3 continues to examine the reactivity of the tin ketyl with labile α-substituents. The reactions of α,β-epoxy ketones with TBTH and their subsequent cyclizations onto unactivated olefins were examined. These reactions, which could form tetrahydrofuran products by addition of oxygen-centered radicals to olefins, were found to undergo an unusual 1,5-tin transfer to yield carbon radicals. These radicals were useful, though, and underwent cyclizations to create highly substituted cyclopentanes.

Once thought to be too unruly for the delicate world of synthetic organic chemistry, free radicals have exploded onto the scene. These intermediates have been generated under mild conditions, have been shown to be very selective, and have tolerated a wide range of functionality. Their viability as a powerful synthetic intermediate has been proven by Dennis Curran and others who have synthesized entire families of natural products with these species. TBTH has shown that it is a valuable synthetic reagent for a variety of transformations. This work attempts to demonstrate that carbonyls can be used effectively as radical precursors with a flexibility that has not been illustrated by either electrochemistry or photochemistry.

CHAPTER 2

CYCLIZATIONS OF α,β-UNSATURATED KETONES

The foundation for this project was established in 1989 by the work of Enholm and Prasad.[63] They studied the cyclizations of O-stannyl ketyls **89** onto both activated and unactivated olefins (Scheme 2-1). The stereoselectivity of the ring closure ranged from 1:1 (anti:syn, relative to the hydroxyl) to 3:1. A natural extension of this work was to attach appendages to the O-stannyl ketyl and see if the radical could be separated from the oxyanion by resonance. The work presented in this chapter was published in a preliminary form as a JACS Communication in 1991.[65]

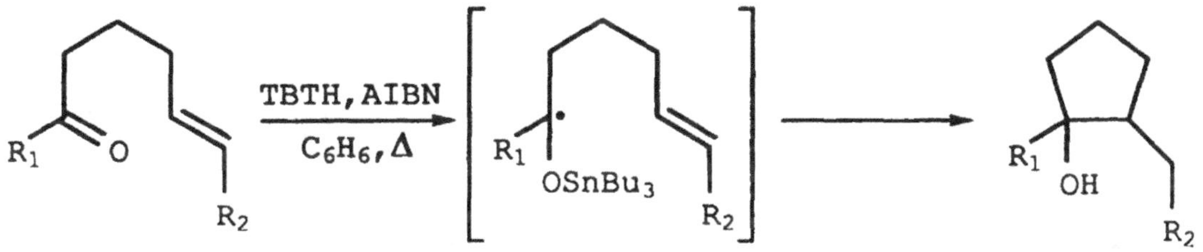

R_1 = Hydrogen or Alkyl
R_2 = Alkyl or EWG

Scheme 2-1

An unsaturated appendage seemed a natural selection due to many factors, which included ease of preparation and effective delocalization through resonance. The general transformation, shown in Figure 2-1, would involve the coupling of the β-carbons of dieneone 91. This is a difficult coupling because the β-carbons in this system both carry a partial positive charge. Prior to our studies there were no reagent-based methods to accomplish this transformation,[66] but there were electrochemical methods.[58,67]

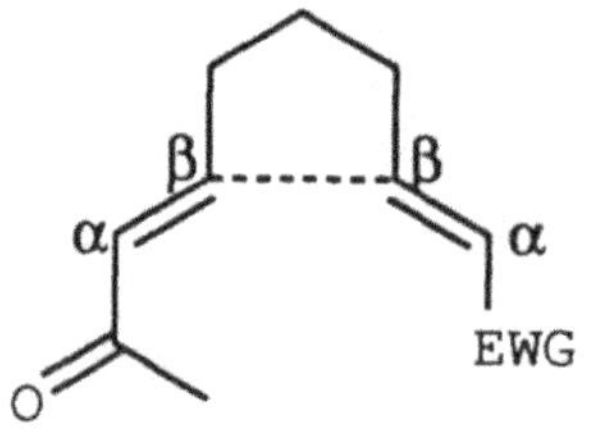

91

Figure 2-1

General transformation of α,β-unsaturated ketones

These methods suffer from some of the basic drawbacks that are associated with electrochemistry. Only conducting solvent systems can be used, and this is usually limited to CH_3CN/H_2O. The conditions need to be closely controlled to avoid unwanted pinacol and aldol products. Aldol and saponification (EWG = CO_2R) products result because basicity builds up at the cathode. Often the specialized equipment is not readily available to a practicing synthetic organic chemist, or it may be expensive. Lastly, this method is not selective enough to be applicable to natural product synthesis which may contain numerous labile functional groups. The stereoselectivities obtained are usually low, 2:1

(anti:syn), although in a single report, Little and Baizer[67a] observed that the addition of $CeCl_3$ improved stereoselectivities (Scheme 2-2).

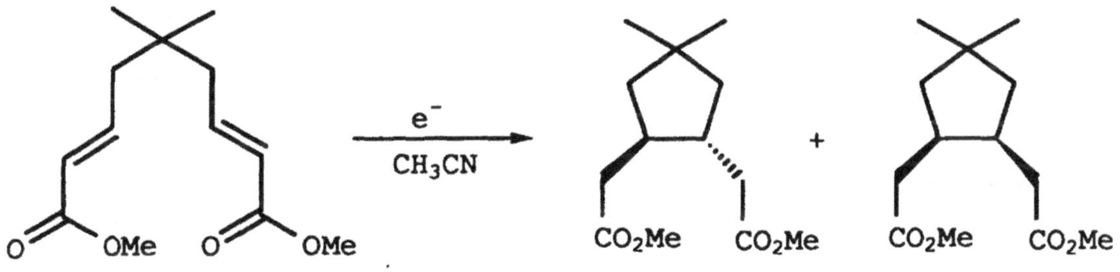

| Without | $CeCl_3$ | 2.6 | to | 1 |
| With | $CeCl_3$ | 15 | to | 1 |

Scheme 2-2

Electrochemists have provided some interesting details on the physical properties of the radical anion ESR spectra of enones, which do not possess acidic hydrogens. These studies show that 50% of the radical density is located on the β-carbon 97 (Figure 2-2), while the remaining portion is divided equally between the carbonyl carbon 96 and the carbonyl oxygen 95.[66a,68]

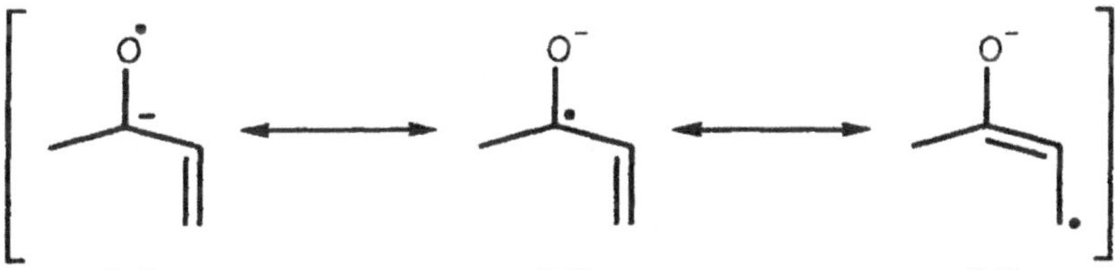

Figure 2-2

Major resonance contributors for the enone radical anion

Our studies were initiated by addressing the question of whether we could obtain an efficient cyclization with substrates such as 88 where the carbonyl is replaced by an α,β-unsaturated ketone to prepare 91. The ketyl, which is produced by the reaction of enone 98 with tributyltin radical, has two major resonance contributors 99 and 100. One might speculate that, if an analogy with the electrochemical resonance structures can be drawn, then 100 should be the major contributor. Also, electronegativity differences between O and Sn should make radical 100 electron-rich. As mentioned before, an electron-rich or nucleophilic radical will prefer to react with an electron-deficient olefin.[1]

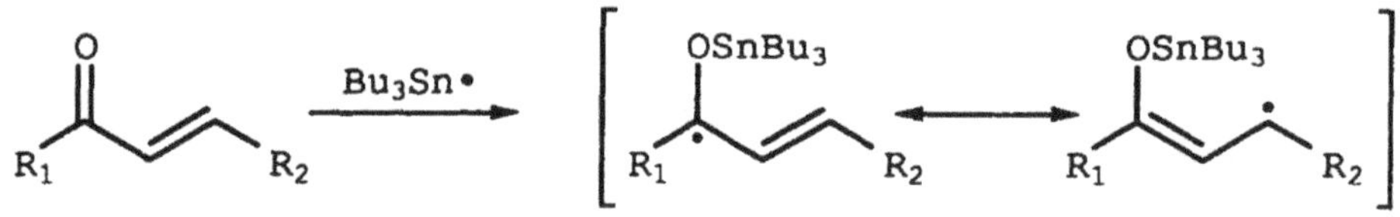

Scheme 2-3

It was first necessary to determine whether activation of the olefin was essential for a successful cyclization. When 101 was subjected to radical cyclization conditions the only product 102 which was observed was a result of simple reduction of the conjugated olefin. These results were not surprising considering that the radical reduction of α,β-unsaturated ketones is a precedented reaction.[69] We had hoped that both activated and unactivated olefins could be used in this study, but this result showed that an activated alkene was a critical element for success.

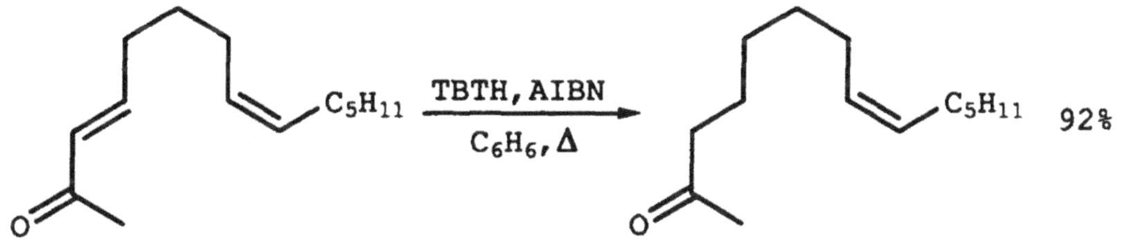

Scheme 2-4

The reason why this cyclization did not work is supported by the frontier molecular orbital theory which was presented in Chapter 1. Enholm and Prasad[63] also found that tin ketyls were reluctant to cyclize with unactivated olefins. The tin alkoxylate in both of the intermediate radicals **24** and **100** impart some negative character onto the radical, which raises the energy of the SOMO. Higher energy SOMO's have better orbital overlap with the LUMO's of electron deficient olefins (Figure 1-2).[2] It is apparent from these cyclizations that the intermediate radical did not effectively cyclize and was quenched by TBTH. Therefore, the rate of cyclization with unactivated alkenes is slow relative to the rate of hydrogen abstraction..

Once it was determined that an activated olefin was an important prerequisite for successful cyclizations, our attention was then focused on the synthesis of the required starting materials. Glutaric dialdehyde **103** seemed ideally suited for the two-step attachment of unsaturated appendages. But, what seemed to be a trivial Wittig reaction to make aldehyde 104 became somewhat of a challenge due to the predominance of the di-addition product. When 1 equivalent of Wittig reagent was added by addition funnel to a large excess (4 eqs.) of dialdehyde the product ratio still favored the bis-product 7:1 (di:mono). This was only a minor inconvenience because the di-addition product 106 was one of the intended starting materials. Eventually, the slow addition of ylide (1 eq.) by addition funnel to 7 eqs. of 103 gave the desired mono-addition product 104 in 74% yield. The second unsaturated appendage was added without problems to yield the needed starting materials (Scheme 2-5).

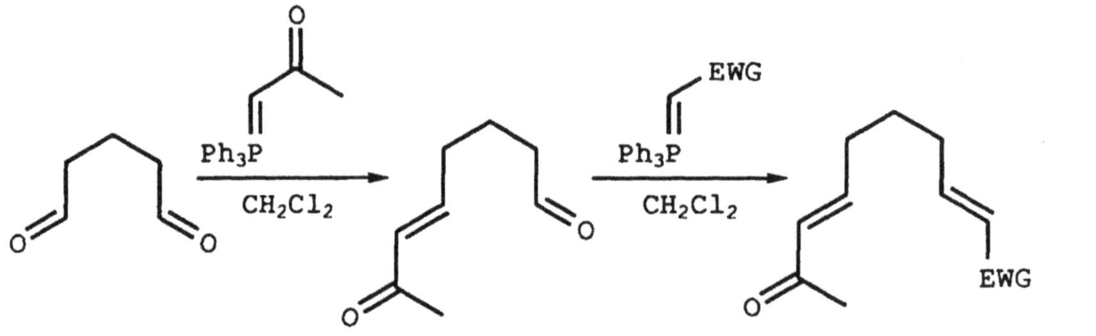

103 **104** 74% **105** **106** **107**

EWG= -CN, 47%, -COMe, 52%, $-CO_2Me$ 49%

Scheme 2-5

In a typical cyclization, the unsaturated ketone was dissolved in benzene (0.1 M), then 3 eqs. of TBTH, and 0.1 eq. of AIBN were added. The mixture was degassed with Argon

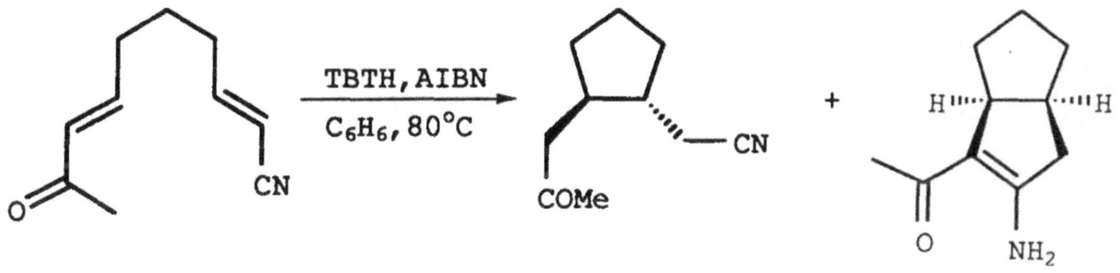

34

Scheme 2-6

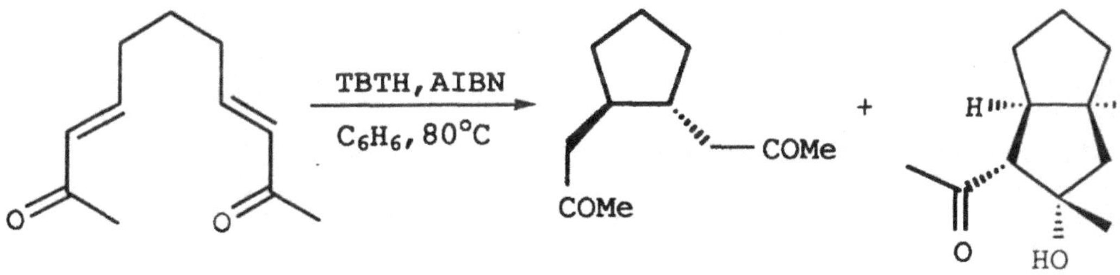

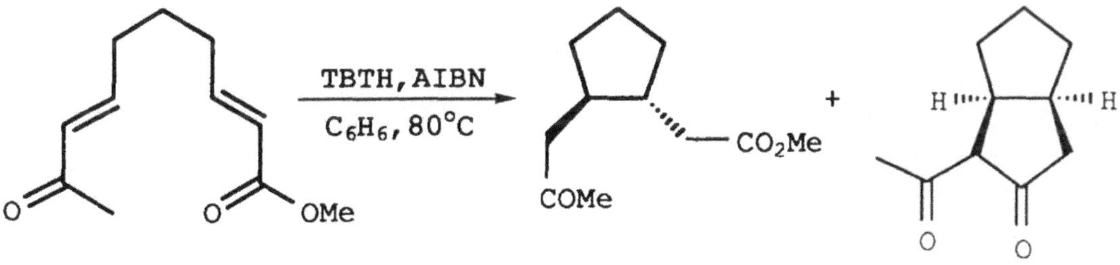

and then heated to 85°C (bath temp.). The reactions were monitored by thin layer chromatography (TLC), and in most reactions the starting material was consumed within two hours. These reactions never seemed to suffer from the sluggish behavior that Beckwith reported,[36] and they were also very clean reactions. A summary of the results of these cyclizations and their products are shown in Scheme 2-6. The ratios and yields reported are from the isolated products and are in good agreement with the GC ratios.

The proposed mechanism for these reactions is presented in Scheme 2-7. The initiation steps which produced tributyltin 11 radical were covered in Chapter 1, Scheme 1-2, and will not discussed further here. Tributyltin radical 11 added to the ketone carbonyl of the starting material to generate the O-stannyl ketyl 115. This ketyl radical, in conjugation with the allylic double bond, has a resonance form with the radical at the β-carbon 116. This resonance form, a 5-hexenyl radical, subsequently cyclized with the activated olefin. The resultant radical species 117 underwent hydrogen atom transfer with a second molecule of TBTH, generating another tributyltin radical which continued the chain process. Tin enolate 118 was then quenched by a proton source to yield the final cyclized product 119.

The minor bicyclic syn products were formed by a second cyclization. The EWG used to activate the olefin could also undergo a second intramolecular reaction with the tin enolate, as in 121. When the appendages were anti to one another they were not close enough to react and the anti-ring fusion in __118__ is too strained,[70] but when they were syn a second aldol-like cyclization occurred (Scheme 2-8). These products demonstrate that the dual reactivity of tin ketyls can be utilized. They undergo an efficient radical reaction followed by a smooth two electron condensation reaction.

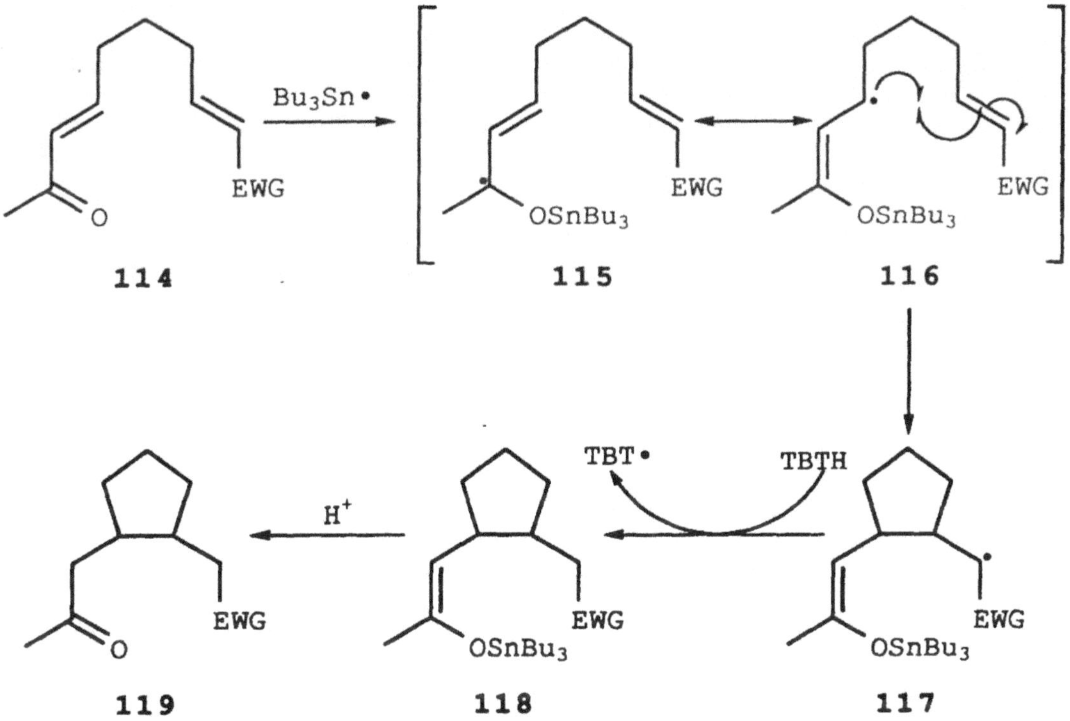

Scheme 2-7

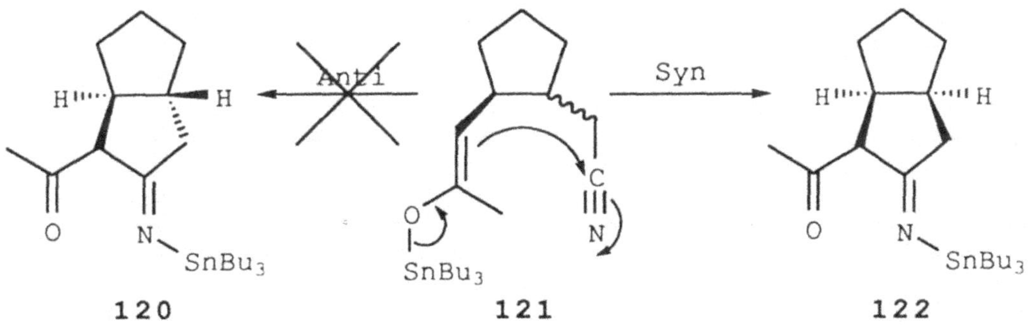

Scheme 2-8

Structural identification of the bicyclo syn products was not always a trivial pursuit. It was believed that the bicyclo syn product 109 from the cyclization of nitrile 105 would be an imine which could be hydrolzed to 113. Unfortunately, this compound resisted attempted hydrolysis with 1 M HCl. After comparison with standard library spectra and conversations with Dr. John Greenhill it was decided that the actual structure was the keto-eneamine shown in Scheme 2-6.

The syn product from the cyclization of ester 107 was a 1,3-diketone. The literature H^1 NMR spectra of 2-acetylcyclopentanone showed contributions from all three tautomeric forms (Scheme 2-9). 123 was the most stable form and has a methyl group at 1.94 ppm. The methyl group of 125 was assigned a chemical shift of 2.21 ppm, and the enolizable proton was barely visible at 3.35 ppm. 113 had a very strong singlet at 2.06 ppm, and a very small singlet at 2.22 ppm, where the ratio is about 10:1 respectively from ^1H NMR integration. This analogy was not strong enough to accurately identify the tautomer, therefore, its structure was unknown.

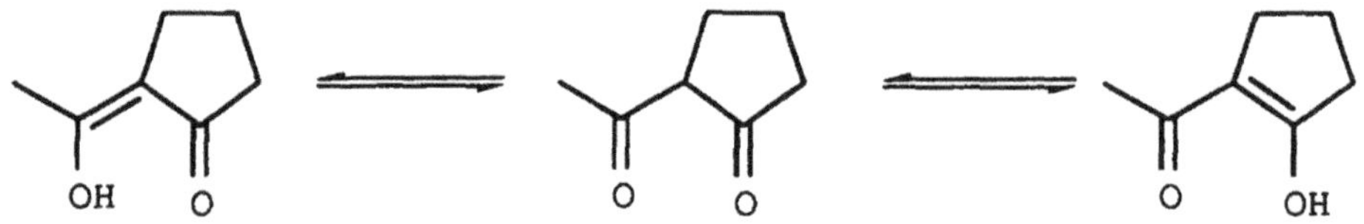

Scheme 2-9

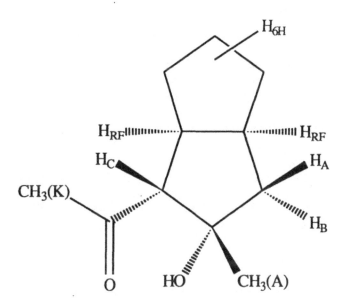

Proton(s) Irradiated	Percent Effect Observed						
	H_A	H_B	H_C	H_{RF}	H_{6H}	$CH_3(A)$	$CH_3(K)$
H_A		30.1	6.0		2.7	2.1 *	
H_B	26.1			6.0			
H_C	2.8				7.8	2.5	*
H_{RF}		1.5			7.2		0.76
H_{6H}							
$CH_3(A)$	*		2.8		2.2		
$CH_3(K)$			*	2.4	3.0^1		

* = Proton difficult to observe due to proximity with irradited proton.
1 = Molecule can adopt a conformation which could give these results.

Figure 2-3

Difference NOE data for <u>111</u>

The syn product 111 from the cyclization of ketone 106 was surprisingly isolated as a single diastereomer. The structure was confirmed by extensive NOE difference studies summarized in Figure 2-3. Additionally, the hydroxyl proton's chemical shift was independent of the concentration of the sample. Therefore, it was believed to be intramolecularly hydrogen bonded to the ketone. It is interesting to note that in this product, four stereogenic centers were created, however only isomer 111 was observed.

Particularly disturbing to us was the anti-stereochemistry of the appendages which stood in marked contrast to nearly all other 5-hexenyl radical cyclizations.[1,2,13] Normal radical cyclizations which produce substituted cyclopentanes conform well to the classic Beckwith model.[40] Beckwith's model (Fig 1-4) suggested that a 1-substituted 5-hexenyl radical gave mostly syn products; our results conflict with his model. An explanation of these deviations are based on several differences between the two radical chair-like systems. Beckwith's system (Scheme 1-8) involved the cyclization of a radical with an alkene trap, and essentially this reaction is kinetically controlled and irreversible. The product ratio was determined by which ever isomer formed the fastest, and was said to be under kinetic control. However, our cyclization involved a resonance stabilized allylic radical reacting with an activated olefin. The resonance stabilization of this radical made the reaction reversible and allowed the less stable syn product to equilibrate and form the more stable anti isomer; this is thermodynamic control. It could also be argued that the high degree of polarity that 105-107 possesses is not adequately represented by the bare skeleton of Beckwith's model.

If our system was indeed under thermodynamic control, then we believed that varying the concentration and the amount of TBTH would lead to improved stereoselectivities. When the reaction was examined at greater dilution, we were delighted to obtain much higher levels of stereoselectivity. Thus, 107 was cyclized at three different dilutions as shown in Scheme 2-10. Increasing levels of anti-stereoselectivity for the ring appendages were obtained as the reaction was diluted. A ratio of greater then 50:1 for the anti:syn products could be achieved at 0.01 M in benzene.

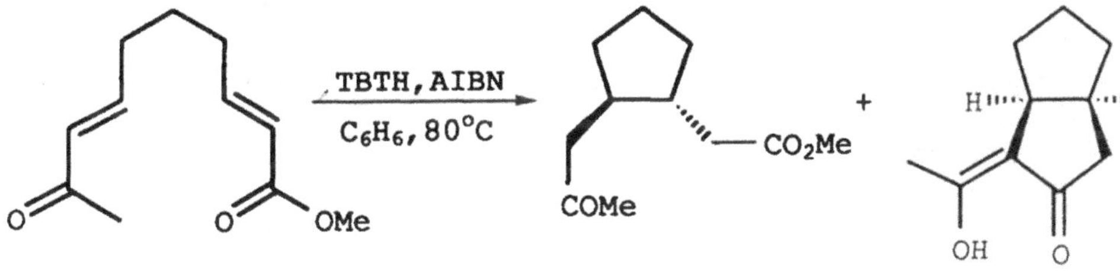

Reaction Conditions	112:113 (% Yield)
1. 1.00 M in benzene, 1.1 eq. TBTH	9:1 (75%)
2. 0.10 M in benzene, 1.1 eq. TBTH	25:1 (82%)
3. 0.01 M in benzene, 1.1 eq. TBTH	>50:1 (82%)

Scheme 2-10

This dramatic increase in anti-stereoselectivity can likely be attributed to the reversibility of the cyclization and the decreased availability of TBTH. Once the cyclized radical 117 reacted with a second molecule of TBTH it was no longer able to open up to 116. Making the reaction more dilute and decreasing the equivalents of TBTH allowed the intermediate radical 117 more time to equilibrate before it was quenched. By the time the radical was quenched the equilibrium would have shifted from the kinetic syn product to the thermodynamic anti product.

It could be that Beckwith's model was applicable to this system, and that with even more concentrated solutions and an excess of TBTH the syn isomer could be the major product. However, it is difficult for a intermolecular quench to compete with an intramolecular equilibrium and it is doubtful that these conditions could be found. The more likely scenario was that Beckwith's model was not entirely applicable to this very polar substrate, and that ratios observed at high concentration may be the kinetic product. Regardless, the dilution study showed that this reaction was reversible and that excellent anti-stereoselectivities could be achieved.

Now that the one-electron chemistry of the tin ketyl had been explored, we attempted to establish the ability to conduct two-electron chemistry as well. Although this has already been demonstrated by the isolation of the minor bicyclic products, two additional experiments were performed

which clearly establish the presence of the stannyl enolate.
Since these reactions were not run in a strictly tandem
sequence, we use the term "serial reactions" to refer to
these studies.   In each experiment, the one-electron
reactivity of an allylic O-stannyl ketyl was induced to
cyclize under the dilute anti-selective conditions (run # 3)
as shown in Scheme 2-10.   The resultant tin enolate was then
immediately quenched in the same pot with Br_2 or D_2O to
produce a *ca.* 2:1 mixture of **126** and **127** as shown in Scheme
2-11.   From ^1H NMR integration of the methylene group α to
the ketone deuterium incorporation was calculated to be
greater then 85%.

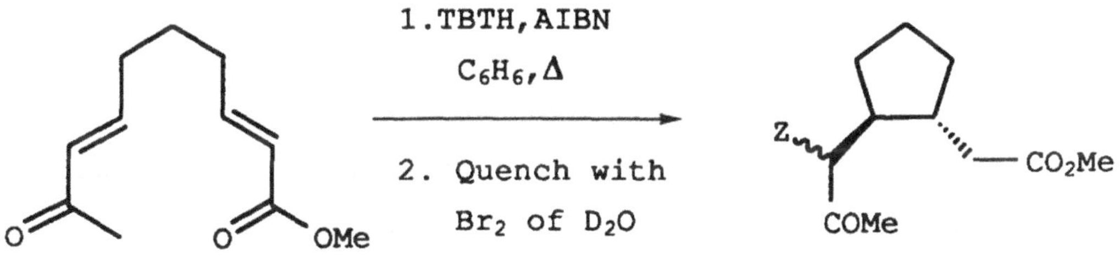

1. TBTH, AIBN
   C_6H_6, Δ

2. Quench with
   Br_2 of D_2O

107

126 Z=Br (86%)

127 Z=D   (87%)

Scheme 2-11

These results and those for the previously discussed
bicyclic products clearly demonstrated the presence and
utility of the tin enolate.   Also the monocyclic products and
dilution studies showed that the radical cyclization of
allylic O-stannyl ketyls could be both efficient and highly
stereoselective.   Collectively these studies show that the

one-electron reactivity in the allylic O-stannyl ketyl can be separated from the two-electron chemistry by sequential transformations, and by the correct choice of experiment. Thus, both types of reactivity can be achieved.

Now that the effectiveness of α,β-unsaturated ketones was demonstrated on an intramolecular level, the next task was to see if it was equally effective on an intermolecular basis. The electrochemistry of α,β-unsaturated ketones has shown that intermolecular coupling is a promising route to highly functionalized cyclopentanes, and they have termed these reactions electrohydrodimerization EHD.[58,67] This name was derived from the overall transformation where the products represented dimerization of the starting material with the addition of two hydrogens.

Often many problems are encountered when one goes from intramolecular reactions, where the course of the reaction can be controlled by the prudent selection of starting materials, to intermolecular reactions which have a great deal of freedom to react along many different pathways. Coupling of unsaturated ketones could occur from both the carbonyl carbon (head) and the β-carbon (tail). This meant that three different coupling products were possible: head to head (pinacol), head to tail, or tail to tail. The hydrodimerization of enone 128 (Scheme 2-12) exemplified the frustrating product mixture which could be produced.[71] Electrochemists believe that these reactions go through either the combination of two radical anions or the addition of a radical anion to the starting enone.[67c]

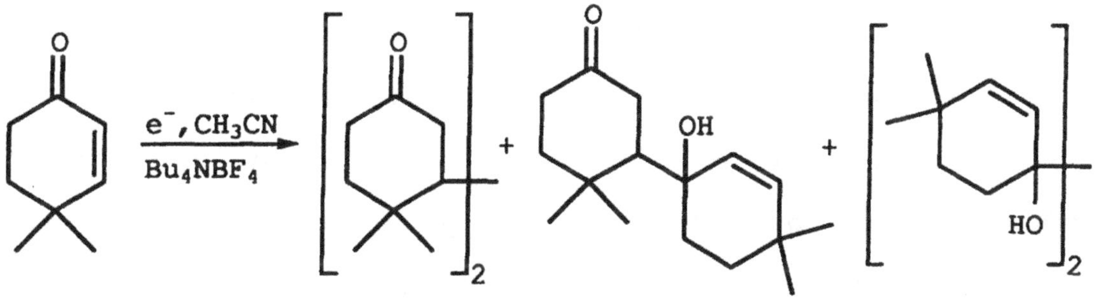

Scheme 2-12

A preliminary attempt to hydrodimerize methyl vinyl ketone under radical conditions produced a low yield of a compound whose ^{1}H NMR spectra resembled an acetylcyclopentanol. This hydrodimerized product further reacted through its tin enolate to give the cyclopentanol. The hydrodimerization of cyclohexenone resulted in the recovery of 37% of the starting material's mass in form of hydrodimerized products. The failure of these reactions to achieve usable yields was most likely due to either the volatility of the starting materials and products or the lifetime of the radical was not long enough to allow for coupling.

Both of these potential problems were circumvented in the hydrodimerization of trans-chalcone 132. The reaction conditions and product ratios are shown in Scheme 2-13.

Notice the delicate balance that existed between the concentration and the equivalents of TBTH. Reaction 1 showed that, if too much TBTH was used, simple reduction of the olefin would be the major product <u>133</u>. In an effort to increase the amount of hydrodimer <u>134</u> that was formed, the reaction was run at even higher concentration with the minimum amount of tin required. We were delighted to find that the previous ratio was reversed and that we were able to isolate hydrodimer <u>134</u> in a 72% yield. It was isolated as an inseparable mix of isomers whose spectra and melting point agreed with published reports.[72]

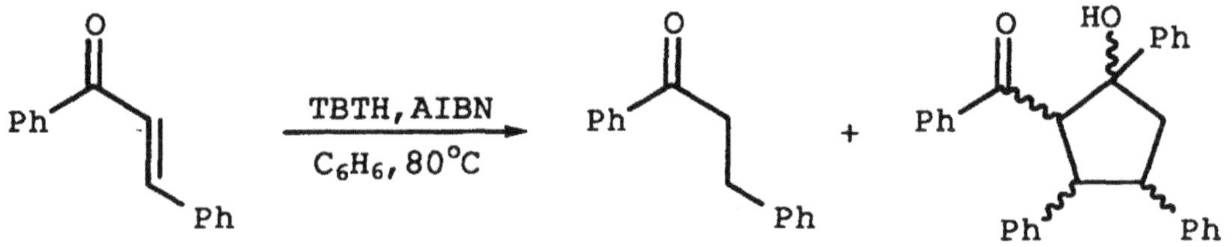

132		**133**	**134**

Reaction Conditions	133:134 (% Yield)
1. 0.5 M in benzene, 1.50 eq. TBTH	4:1 (93%)
2. 1.0 M in benzene, 0.55 eq. TBTH	1:3 (94%)

Scheme 2-13

A proposed mechanism for the hydrodimerization of trans-chalcone is shown in Scheme 2-14. <u>134</u> was formed by the initial β—β coupling of the allylic O-stannyl ketyl <u>135</u> with a molecule of trans chalcone, and resultant radical __136__ was quenched with TBTH to give tin enolate __137__. This enolate then condenses intramolecularly on the ketone to give tin alkoxide __138__ that upon hydrolysis yielded the final hydrodimer __134__.

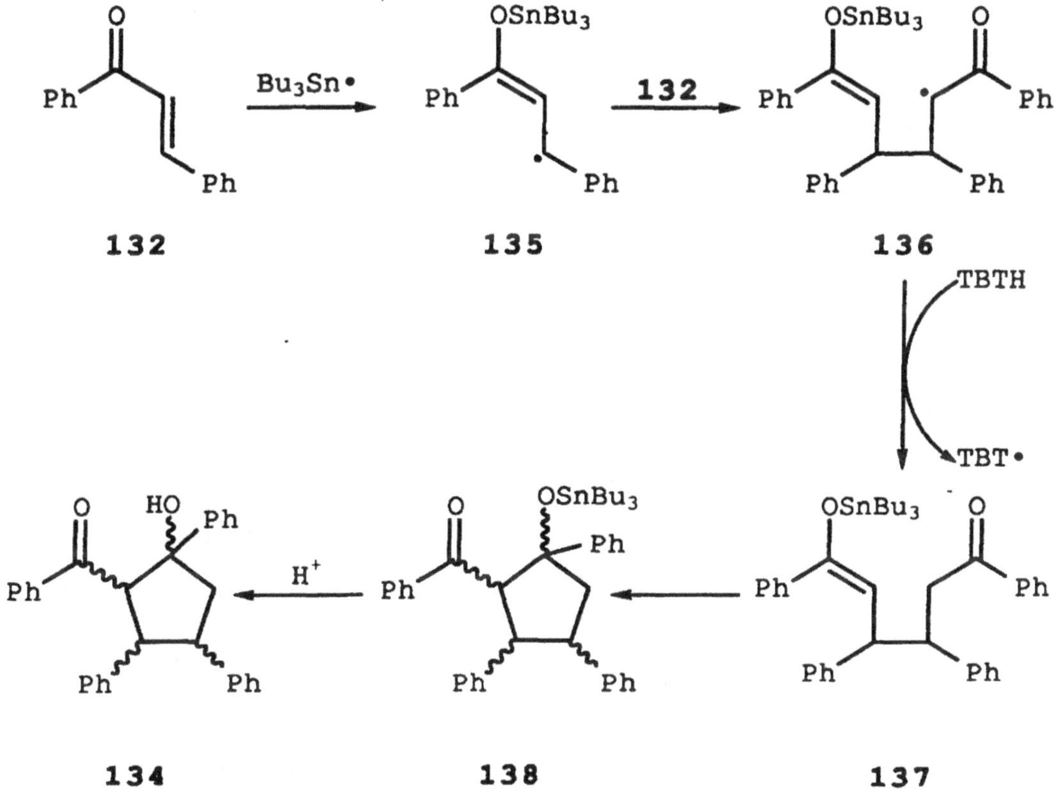

Scheme 2-14

Although hydrodimers were achieved by this method it was believed that the scope of this reaction was not sufficiently broad to make it synthetically useful. Stabilization of the allylic O-stannyl ketyl radical with phenyl groups seemed to be imperative to the success of this reaction. The tentative balance between reaction concentration and TBTH concentration also limited this reaction. This problem could be avoided by the use of either hydrideless tin sources or syringe pump techniques. Both of these methods keep the concentration of hydride donor at very low levels, but if these reactions were run at high reaction concentration and low hydride density then the most likely products would be polymers.

In conclusion, the intramolecular coupling of α,β-unsaturated ketones to activated olefins can be applied to the synthesis of substituted cyclopentanes. Excellent anti stereoselectivities can be achieved with the proper selection of reaction conditions. Additionally, the dual reactivity of the allylic O-stannyl ketyl radical can be separated and utilized in sequential one- and two-electron reactions. Although, the intermolecular coupling of unsaturated ketones was demonstrated, additional work is needed for this methodology to be synthetically useful. This additional work should be focused on hydrideless tin and syringe pump techniques, also the prospect of cross-coupling reactions should be investigated. Collectively, this work has illustrated that α,β-unsaturated ketones are a viable radical precursor, and the degree of functionality and stereocontrol which can be achieved allows this methodology to be applied in the synthesis of natural products.

CHAPTER 3

CYCLIZATIONS OF α,β-EPOXY CARBONYLS

In continuation of our studies directed towards the use of carbonyls as free radical precursors, we examined the free radical ring-opening reactions of epoxy ketones and aldehydes. The radical ring opening of cyclopropanes is known to be an extremely rapid process,[73] likewise, the analogous opening of epoxides is thought to be equally rapid.[2] The main difference between these two system is the radicals which can be produced. The cyclopropyl case can yield only carbon-centered radicals, while the epoxide example can afford either oxygen-centered or carbon-centered radicals, depending on which bond of the epoxide fragments.

The direction of fragmentation is largely controlled by the nature of the R_1 substituent (Scheme 3-1). Jorgensen[74] has suggested that the bond dissociation energy of the C-O bond is lower than the C-C bond except when stabilizing groups are present to lower the energy of the C-C bond. When this is a radical-stabilizing group such as a vinyl or aryl group, then fragmentation of the C-C bond leads to a low energy and delocalized carbon-centered radical. When no stabilizing groups are present, cleavage of the weaker C-O bond will be the predominant reaction. Synthetic methods have been developed which utilize both of these fragmentation pathways.

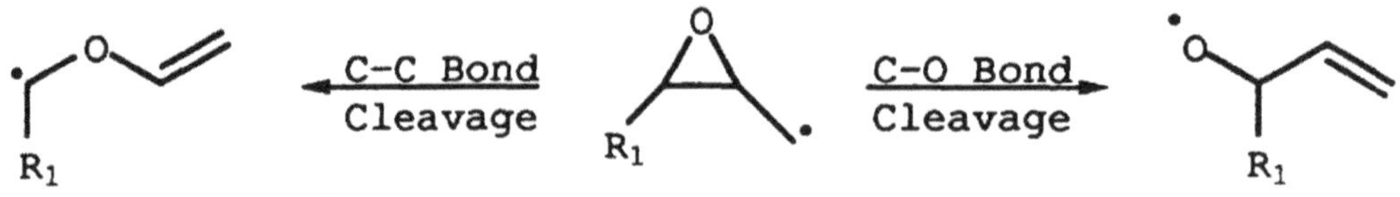

Scheme 3-1

The route which has received the least attention has been the fragmentation of the carbon-carbon bond to give stabilized radical 141. The reason why there has been less research focused on this methodology is because the requirement of a radical stabilizing appendage tends to limit its applicability. Additionally, some researchers have found that the presence of either vinyl[75] or aryl[76] substituents do not guarantee the formation of carbon-centered radicals. But, it is generally accepted that the presence of these groups will lead to cleavage of the C-C bond.[77]

Feldman and Fisher[75] used this fragmentation to generate functionalized THFs. Radical 147 was generated by addition of phenylthio radical to epoxyalkene 144. The epoxide ring was opened by cleavage of the C-C bond to give stabilized radical 148, which reacted with the activated olefin to give adduct 149. This radical then cyclized onto the olefin which was created by the epoxide opening, and the final product was produced by elimination of thio phenyl radical.

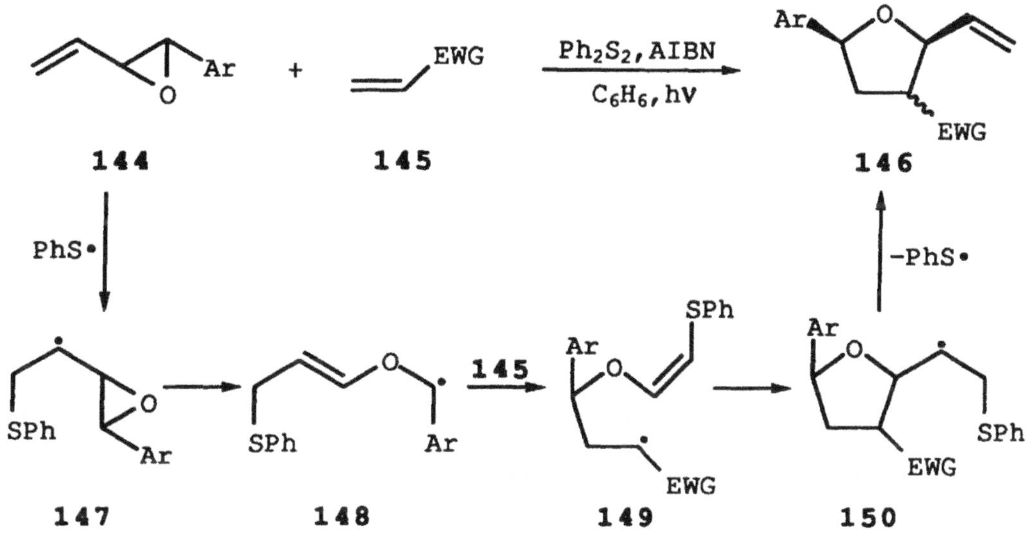

Scheme 3-2

The syn-2,5 products were the only diastereomers which were observed, and this conformed to the stereochemistry which was predicted by Beckwith's model (Figure 1-4). Very little control at the C-4 position was demonstrated, but this can be blamed on the epimerizability of this center and not necessarily on the cyclization. This reaction represents a formal [3 + 2] addition where essentially both electrons of a carbon-carbon bond are used to make the new bonds. This is a very interesting approach to the synthesis of THFs and demonstrates that all retrosynthetic disconnections should be explored.

The majority of the classical work in the late 1960s and early 1970s dealt with the opening of epoxides, mostly concerning the pathway which produced oxygen-centered radical 143. In these papers, there are a wide variety of methods

51

for the production of the α radical, and many different synthetic and mechanistic goals. The common thread which runs through these papers is that a radical is generated α to an epoxide which causes the epoxide ring to open, and the resultant alkoxy radical helps accomplish their transformation. Some of this work, which pertains to this dissertation, will be reviewed here.

If radical 143 was simply quenched by hydrogen atom transfer with TBTH then allylic alcohols would be the product, and many papers have utilized the fragmentation in this manner.[78] This transformation could be thought of as a convenient alternative to the Wharton reaction,[79] where epoxy ketones were reacted with hydrazine to give allylic alcohols. It was discovered in 1971 by H. C. Brown and coworkers[78a] when alkyl radicals from organoboranes added to vinyl epoxides to form allylic alcohols.

More recently in 1981, Barton and coworkers[78b] found that product formation was highly dependent on the reaction conditions. The fragmentation of epoxide 151 under "normal addition" conditions did not give the expected allylic alcohol as the major product, but instead, gave ketone 153 which resulted from an alkyl migration caused by β-scission of the C–C bond. "Normal addition"[80] refers to TBTH and AIBN being added dropwise to a refluxing solution of starting material, which effectively maintains a low concentration of hydride donor. This low concentration of hydride donor allows the intermediate radicals to do other chemistry.

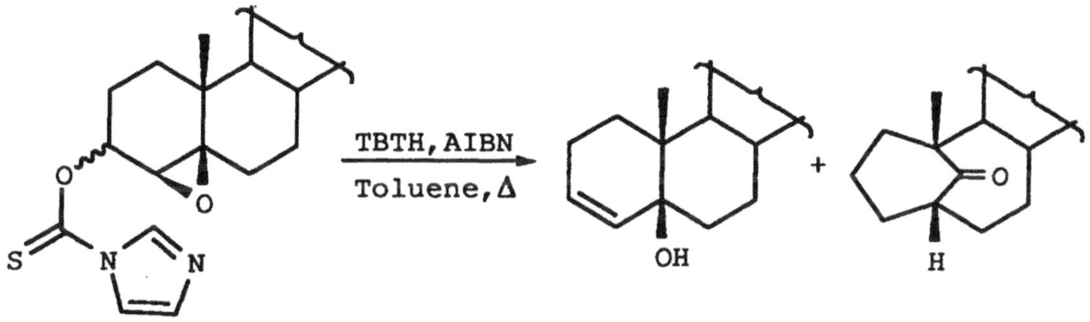

151   **152**   **153**

<u>Reaction Conditions</u>   **152:153** (% yield)

1. 2 eqs. TBTH, Normal Addn.   1 : 2.0 (71%)

2. 9 eqs. TBTH, Inverse Addn.   4.3 : 1 (72%)

Scheme 3-3

Barton and coworkers felt that, if they could maintain a very high concentration of hydride donor, they would be able to isolate the allylic alcohol in useful yields. The "inverse addition" method that they used maintained a very high concentration of TBTH by dropwise addition of the starting material to a refluxing solution of TBTH (9 eqs.) and AIBN. The high concentration of hydride donor quenched the intermediate alkoxy radical before it could do other chemistry. Other examples in this paper showed that "inverse addition" helped avoid unwanted side reaction products which were formed when the "normal addition" method was used. However, these unwanted products showed that alkoxy radicals can do a great deal more then simple hydrogen abstraction, and groups began to explore their potential.

Johns and Murphy[81] were intrigued by some of the products which were described in Barton's paper and set out to see if the alkoxy radical could be used in the synthesis of tetrahydrofurans. Starting material 154 is easily prepared from commercially available geranial. To keep the concentration of TBTH at a low level they used the "normal addition" process and isolated THF products in pretty good yields (Scheme 3-4). Although no ratios were given, they mentioned that in the R=n-butyl case, the major diastereomer had an anti-relationship between the vinyl and the isopropyl group, which was in agreement with the major products formed in Beckwith's studies (Scheme 1-8). Interestingly, bicyclic product 156 was formed by a tandem radical cyclization where the radical formed after the first cyclization reacted with the vinyl group in a second cyclization.

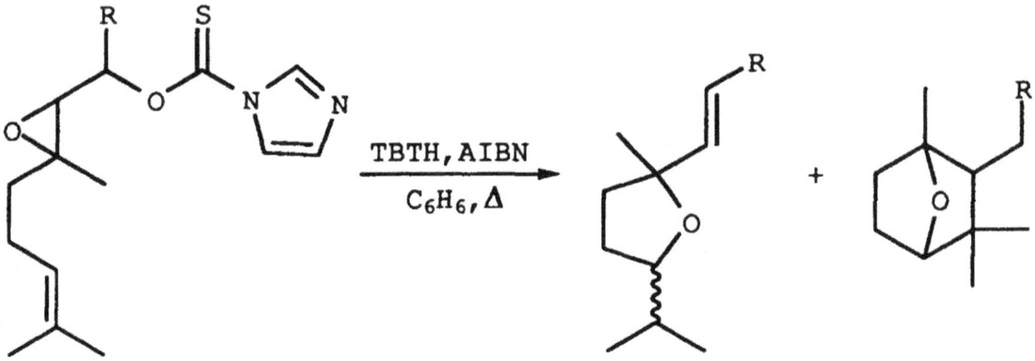

	154		155	156
R = Me			53%	14%
R = n-Bu			63%	22%

Scheme 3-4

Alkoxy radicals are well known for their ability to abstract hydrogen atoms through a six-membered transition state from neighboring atoms which bear hydrogens, which was demonstrated in the classic Barton reaction.[82] There was no mention of any products derived from this pathway in the reduction of 154. The only hydrogens which could have been abstracted through a six-membered transition state were next to the olefin, and these were not accessible due to the trans geometry of the olefin. However, some studies have used this prominent reaction of alkoxy radicals to generate polycyclic compounds.

Rawal and coworkers[54] used the hydrogen-abstracting ability of alkoxy radicals to facilitate the cyclization shown in Scheme 3-5. Addition of tributyltin radical to the thiocarbonyl of 157 caused it to decompose to radical 160. The epoxide ring was then cleaved by this radical to produce alkoxy radical 161, which then abstracted a hydrogen atom from the δ-carbon. The resulting stabilized radical 162 was then in position to cyclize onto the olefin produced from the fragmentation of the epoxide.

Bicyclic products were formed in a 69 % yield, and the ratio between the diastereomers 158:159 was 2.7:1. Product 159, which has the ester pointing into the cup of this cis-fused bicyclic, was epimerized to the more stable 158 by treatment with catalytic amounts of t-BuOK. This method also produced successful cyclizations on other cyclic and acyclic

systems.   Good results were also achieved when alkyl
appendages, with no radical-stabilizing groups, were used.

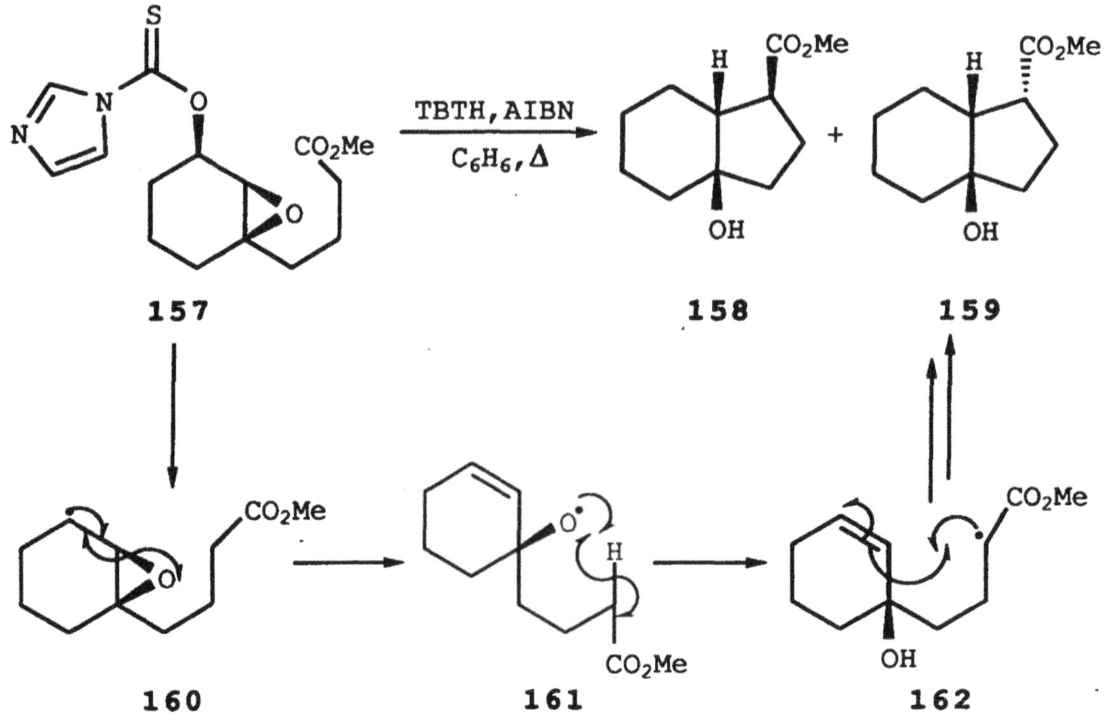

Scheme 3-5

Kim and coworkers[83] used the addition of tributyltin
radical to vinyl epoxides in much the same way. The addition
of tin radical to the vinyl group of 163 provided the radical
which opened up the epoxide, which then abstracts a hydrogen
from the δ-carbon. This radical then cyclized back onto the
olefin which was created when the epoxide was fragmented, and
the final product is formed by ejection of tributyltin
radical. Kim's group also had success with alkyl appendages
that contained both electron withdrawing and electron
releasing substituents.

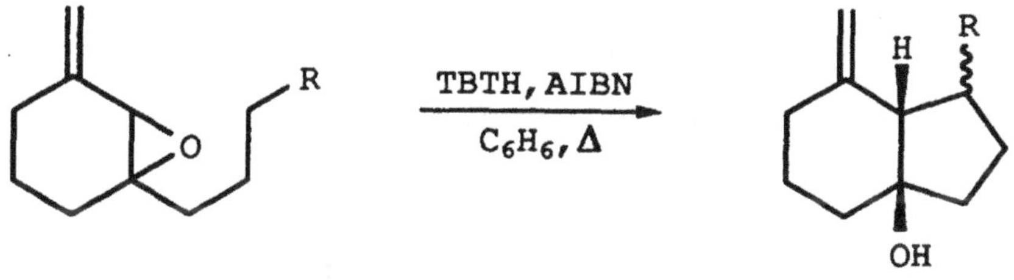

163 **164**

Scheme 3-6

In the same paper, Kim and coworkers reported similar cyclizations which occurred through a rare 1,5-tin transfer. This mechanism was first proposed by Davies and Tse[84] who did ESR spectroscopy on the reactions of glycidyl tributyltin ethers <u>165</u> with t-butoxy radicals. These studies found that the predominant radical in solution was <u>168</u>. The authors reported that this was the first time that a 1,5-organometallic transfer had been observed. This mechanism seemed suspicious at first, but we now believe that our reactions are occurring by the same mechanism, which will be discussed later.

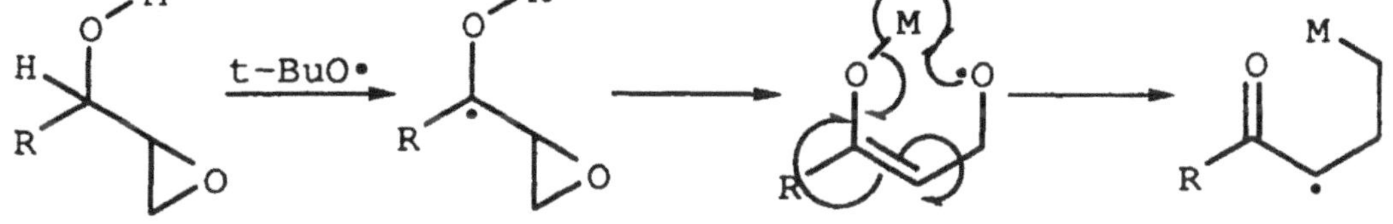

165 **166** **167** **168**

M = $SnBu_3$
R = H or alkyl

Scheme 3-7

Other mechanistic studies have looked at the radical-induced opening of epoxides from another angle. These papers were more concerned with the stereoelectronic factors which governed the opening of epoxides. Bowman and coworkers[85] used the norborane skeleton as a template for their reactions. They envisioned that the exo-oxiranes 169 would lead to C-O bond cleavage products 171, while the endo-oxiranes 172 would produce C-C bond fragmentation products 174. This prediction was based on the stereoelectronic control observed for the radical ring opening of analogous cyclopropyl and cyclobutyl systems.[86] It was proposed that the transition state required maximum overlap of the radical's SOMO with the σ-bond which was undergoing scission.

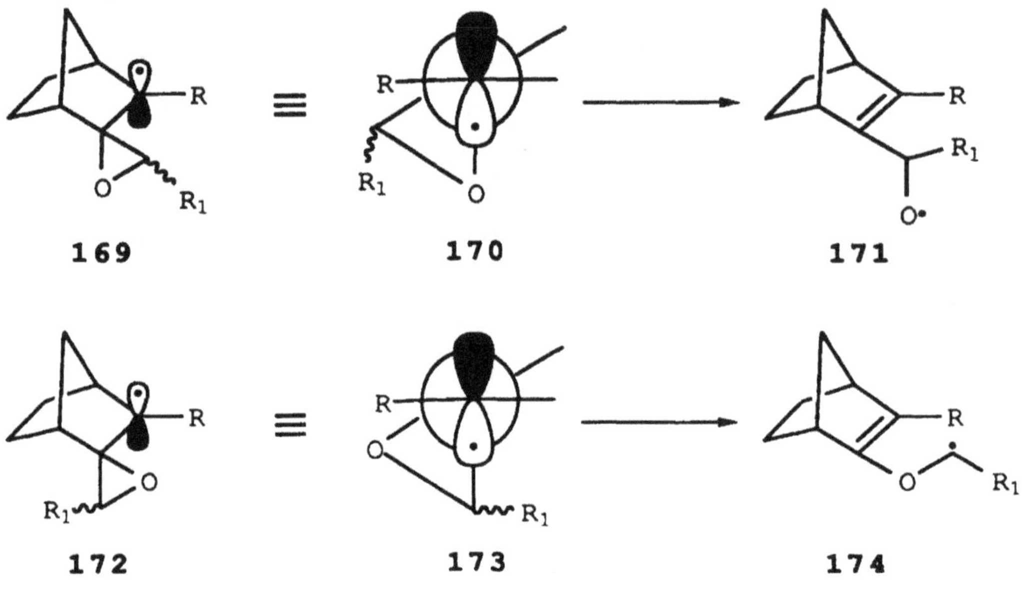

169   170   171

172   173   174

R = H or OSnBu$_3$
R$_1$ = H or Aryl

Figure 3-1

Stereoelectronic model for Exo- and Endo-epoxides

Bowman and coworkers envisioned this system as the perfect test for these stereoelectronic factors. Many of the current papers dealing with epoxide openings invoke this as a potential mechanism, but it had never been tested. However, Bowman found that no stereoelectronic effect could be observed. When aryl derivatives of 169 and 172 (R=H and R$_1$=Aryl) were generated from the corresponding bromides, no products from the cleavage of the C-O bond could be isolated. Additionally, when non-aryl derivatives of 169 and 172 (R=OSnBu$_3$ and R$_1$=H) were created from the corresponding ketones, no products from the cleavage of the C-C bond could be found. They concluded that no stereoelectronic effects were operational, and that the determining factor was the strengths of C-C and C-O bonds.

Many of the twists and turns which were observed in the works reviewed above, also found their way into the study which is presented in this chapter. Many of these diversions were fully anticipated because of the well-known ability of alkoxy radicals to abstract hydrogen atoms. However, what was not anticipated was that the rare 1,5-tin transfer mechanism would be the proposed mechanism. This project started out as an extension of my oral research proposal, which presented the idea of synthesizing THFs, such as 175, through the fragmentations of epoxy ketones. However, the fragmentation of epoxy ketone 176 which was used in this project ended up yielding cyclopentanols 177.

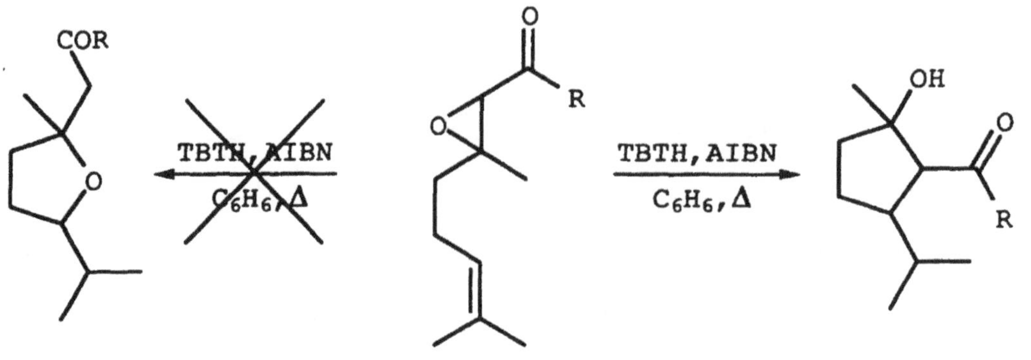

R = H, Me, n-Bu

Scheme 3-8

The molecule which is shown in Scheme 3-8 was very similar to the system which Johns and Murphy[81] used to synthesize tetrahydrofurans in Scheme 3-4. The only difference between the studies was that they used thiocarbonyls to generate the radical and we used ketones. One would expect that minor differences, such as the generation of the initial radical, would have little effect on the outcome of the reaction. However, so often when you least expect something unusual, that is exactly what you get. Anyway, we embarked on the study of the free-radical fragmentation of epoxy ketones, with the hope of developing methodology related to their cyclizations.

The synthesis of the starting materials for this study were prepared from geranial, which either could be purchased or was easily prepared by oxidation of geraniol, and is shown in Scheme 3-9. The geraniol which was commercially available was a mixture of cis and trans isomers in about a 1:2 ratio, respectively. No attempt was made either to separate them or to examine their individual reactivity.

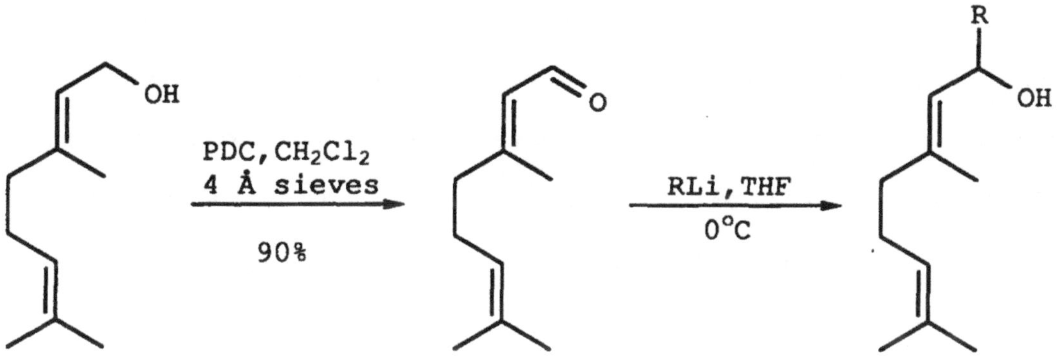

177 **178** R = Me **179** 95%

= nBu **180** 96%

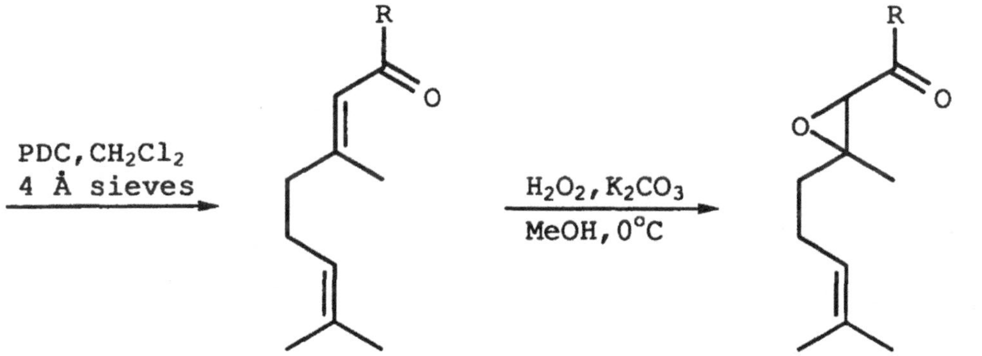

R = Me **182** 71%

= nBu **183** 39%

R = H **185** 27%

= Me **186** 80%

= nBu **187** 98%

Scheme 3-9

The allylic oxidation of geraniol 177 proceeded smoothly to produce geranial 178 in 90% yield. The 1,2 addition of organolithiums to geranial was carried out at 0° C in THF and gave allylic alcohols 179 and 180 in very good yields. No products from the 1,4 addition of the organolithium were observed. The second series of allylic oxidations with PDC did not proceed as smoothly, but gave unsaturated ketones 182 and 183 in reasonable yields. The alcohol-directed epoxidation of the allylic alcohols was not attempted because in a similar case that we had tried it was found that the other olefin was sometimes epoxidized. So the epoxidation of the unsaturated ketones was attempted by using a 30% aq. hydrogen peroxide solution and K_2CO_3 as the base. This protocol prepared the desired starting materials with overall yields of 24% for 185 (R=H, 2 steps), 49% for 186 (R=Me, 4 steps), 33% for 187 (R=n-Bu, 4 steps).

In a typical cyclization, the epoxy ketone was dissolved in benzene (0.5 M), then 2.5 eqs. of TBTH, and 0.1 eq. of AIBN were added. This mixture was degassed with argon for 20 minutes, then the reaction was heated to 85° C (bath temp.). The reactions were monitored by TLC, and in most cases the starting material was consumed within a few hours. These reactions were run at higher concentrations because at first it seemed that the epoxide's opening may have been sluggish, but epoxide openings were also achieved at low concentrations, such as 0.025 M for 185.

The conditions that the reactions were run at and the ratio of products are shown in Scheme 3-10. The precursor was used several times because it was easily accessible from commercially available materials. However, the drawback to this precursor was that the products which were formed were generally a mixture of isomers due to four appendages on the products. This mixture was not easily separated, and a thorough identification of stereochemistry was not considered possible. In general, it was believed that there were only two isomers formed. This was less then might be expected with three different chiral centers in the product, which could give rise to four different diastereomers.

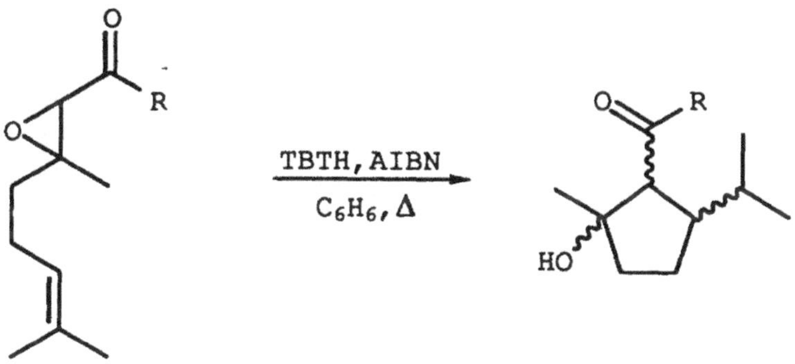

Substrate	Conditions	Major:Minor (% Yield)
186 (R=Me)	0.5 M, 2.5 eq TBTH	**189:190** 3:2 (83% Yield)
187 (R=nBu)	0.5 M, 2.5 eq TBTH	**191:192** 1:1 (81% Yield)

Scheme 3-10

The proposed mechanism at this point is shown in Scheme 3-11 for the methyl example. We believed that the products shown in Scheme 3-10 were a result of the proficient ability of alkoxy radicals to abstract hydrogens, such as in 194. This would generate allylic-stabilized radical 195 which also can be represented as resonance form 196. The 5-hexenyl cyclization appeared to proceed smoothly yielding 197, which, after additional reactions with TBTH and a proton source, would yield the products shown in Scheme 3-10.

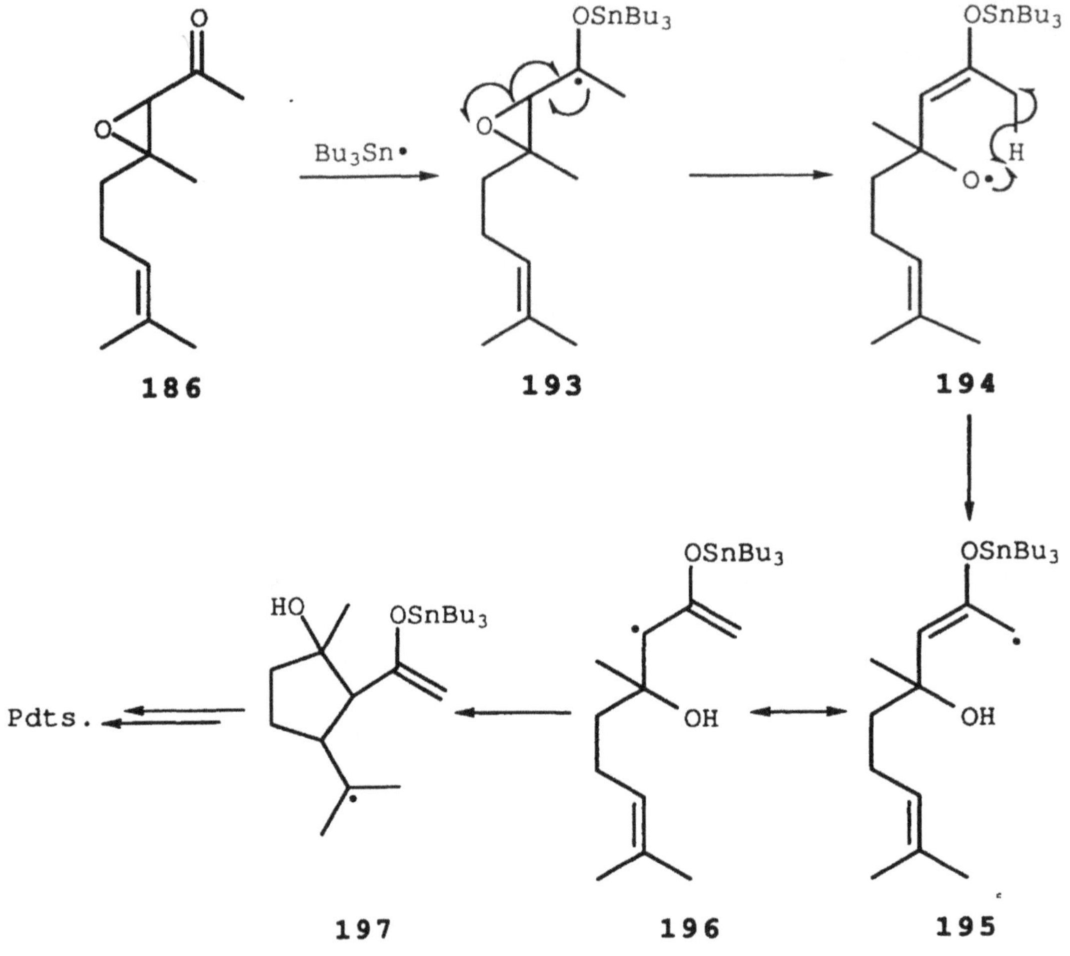

Scheme 3-11

For this mechanism to be operative there needed to be hydrogens on the group which was attached to the ketone, as in __194__ in Scheme 3-11. If there were no hydrogens, then radical __195__ could never be generated. Compound __185__ possessed no abstractable hydrogens, and so its cyclization, or inability to cyclize, would be a key test for this proposed mechanism. When this reaction (Scheme 3-12) was carried out we were surprised to find that __185__ cyclized efficiently to afford cyclopentanols.

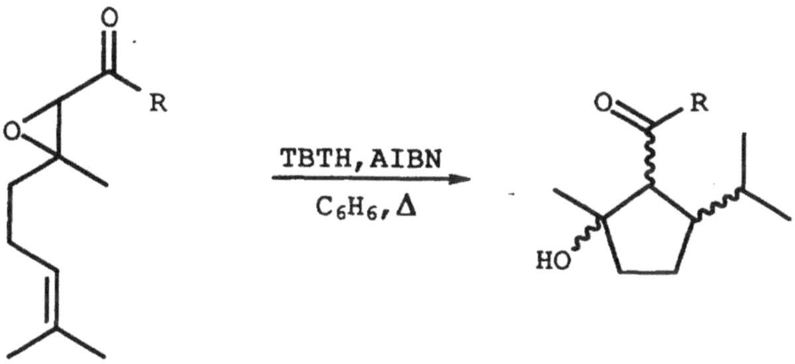

Substrate	Conditions	Major:Minor (% Yield)
__185__ (R=H)	0.1 M, 2.0 eq TBTH	__198:199__ 3:1 (81% Yield)

Scheme 3-12

At this point we realized the original mechanism that was proposed (Scheme 3-11) was not possible. After careful examination of the literature, the references to 1,5-tin transfer (Scheme 3-7) were found,[76b,83,84] which at first

65

seemed to be a bit obscure, but in time we realized that this was indeed the operative mechanism. The proposed mechanism, now revised as it applies to the cyclization of 185, is shown in Scheme 3-13.

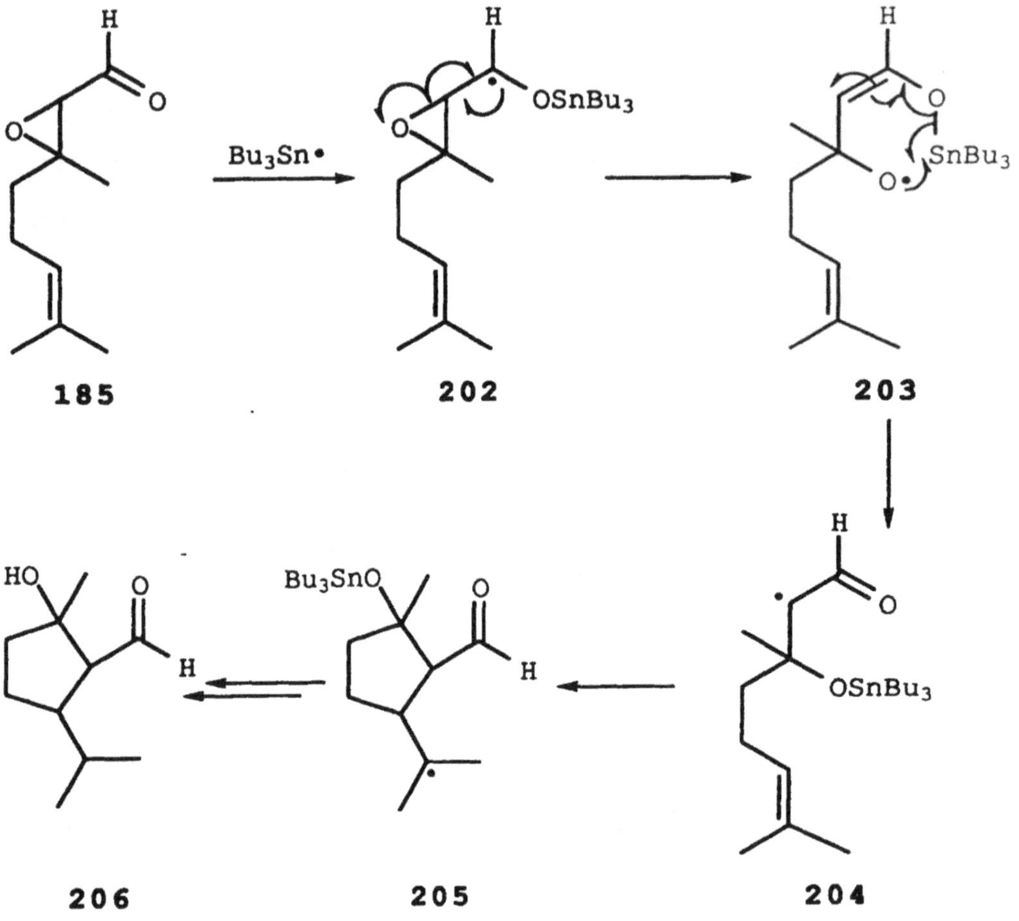

Scheme 3-13

The deciding factor between these mechanisms was the geometry of the olefin which was produced when the epoxide was fragmented (Figure 3-2). If the alkoxy radical and the tin enolate are cis to one another, as in 209, then 1,5-migration can occur; but, if they are trans, then the

migration cannot occur. If a ketone bears a group with α
hydrogens then the alkoxy radical in 210 could abstract them
and the reaction could proceed by the mechanism in Scheme 3-
11.

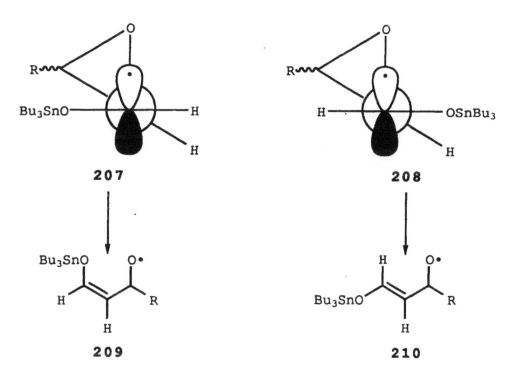

Figure 3-2

Transition states model of cis and trans tin enolate

So the dilemma we were faced with was that there were
two possible mechanisms which the reactions discussed in this
chapter could follow. The initial reactions in Scheme 3-10
could go by either mechanism, while the reactions in Scheme
3-12 most likely followed the 1,5-tin transfer mechanism.
The tin transfer mechanism must proceed through transition
state 207 which is more sterically congested. We believe
that a weak chelation effect between the tin enolate and the

resulting alkoxy radical would help overcome any steric problems. This six-membered ring chelate would eventually develop into the transition state for the 1,5-tin transfer. A new system was devised which would illuminate the reaction pathway of the initial reactions in Scheme 3-10. The synthesis of this compound is shown in Scheme 3-14.

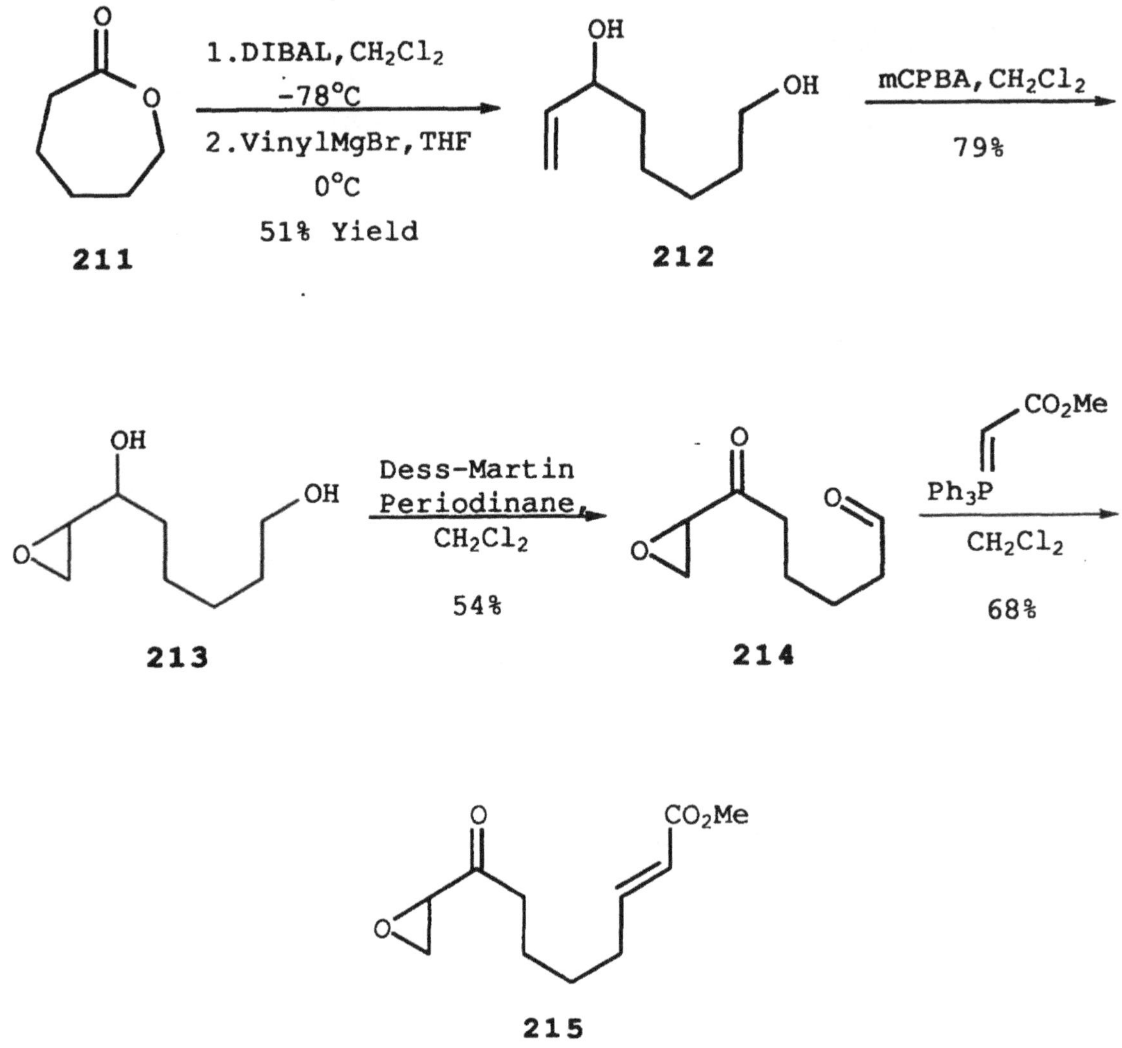

Scheme 3-14

The low temperature reduction of ε-caprolactone with DIBAL produced a product which seemed to polymerize rapidly, and was not characterized. Instead, after a crude purification, it was treated with vinyl magnesium bromide which yielded diol **212** in 51% yield for the two step reaction. This diol was then epoxidized using mCPBA to give a very polar product which was extremely hard to chromatograph. PDC oxidation of diol **213** did not produce a large amount of the desired product. This might have been because the highly polar substrate was difficult to separate from the PDC. The highly touted Dess-Martin Periodinane[87] was synthesized, and 2 eqs. of it was used for the oxidation which proceeded smoothly to produce **214** in a 68% yield. The final product was synthesized by a stabilized Wittig reaction on the aldehyde to give **215** in 10% overall yield in five steps from caprolactone.

The reason the cyclization of **215** was important was because it could help provide additional information about whether the hydrogen abstraction mechanism is valid for these fragmentations. If this cyclization succeeded in producing cyclopentanes, then the alkoxy radical must be abstracting a hydrogen; if it does not succeed, then the 1,5-tin transfer mechanism must be valid. The starting material was dissolved in benzene (0.01 M), then 1.5 eqs of TBTH and 0.1 eq. of AIBN were added. The reaction was degassed with argon and the reaction was refluxed. The only product which was isolated

was the opened epoxide 216. No cyclized products could be
isolated.

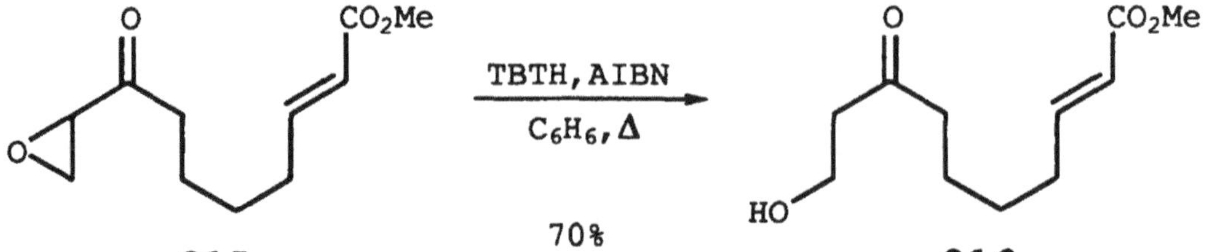

Scheme 3-15

This suggested that the mechanism in Scheme 3-10, or the
hydrogen abstraction mechanism, was a less-likely
possibility. We felt that hydrogen abstraction by the alkoxy
radical would generate a 5-hexenyl radical which would likely
cyclize onto the activated olefin. It should be noted that
hydrogen abstraction may have actually occurred but the
ketone stabilized radical was electron poor and could not add
to the LUMO of the alkene The probable mechanism which can
now be proposed is the 1,5-tin transfer which was originally
proposed by Davies.[84] When this tin transfer occurred in the
reaction of 215 a radical was generated α to the ketone, but
too far away to cyclize with the activated olefin, and it was
simply quenched with TBTH to give the final product.

In conclusion, the work which is presented in this
chapter has shown that the fragmentation of epoxy ketones
proceeds by a relatively novel 1,5-tin transfer. The
products of these reactions are highly functionalized
cyclopentanols. Although the stereochemistry of this

cyclization could be improved, the number of isomers isolated was far less then might be expected from such a densely functionalized molecule. In this regard, radical cyclizations were conducted on <u>186</u> and <u>187</u> with SmI$_2$ and the same products as mentioned above were isolated. However, initial TLC showed that a much higher degree of stereoselection (around 10:1) was possible. But the ketone center turned out to be very epimerizable, and these selectivities were never realized. Future work in this direction could potentially yield a very a fruitful study.

CHAPTER 4

SUMMARY

The studies described in this dissertation are an attempt to expand the realm of free-radical chemistry. Free-radical methodology has been dominated by halogen and group abstraction as the source of organic radicals. Although, these methods have helped establish free-radical reactions as a valuable tool in the synthesis of complex molecules, hopefully, this thesis will show that there are quality alternatives to these methods. The mild reaction conditions, and the ability to control reactivity, stereoselectivity, and regioselectivity in these reactions are certainly different qualities from the original "unruly" reputation of free radicals 40 years ago.

A problem associated with "classical" radical chemistry is illustrated by the work of Dennis Curran shown in Scheme 1-1. In his synthesis of hirsutene, Curran took a molecule which had three exploitable functional groups (1) and produced a molecule which had only functional group (2). The use of free-radical cyclizations on this molecule had virtually stripped it of usable functionality. Although, the loss of functionality was intended in the case of hirsutene, it still demonstrates that sometimes radical cyclizations

71

result in compounds which can not be further functionalized. A goal of this thesis was to show that radical cyclizations could be conducted successfully with carbonyls as free-radical precursors.

In the first study, TBTH was used to couple the β-carbons of α,β-unsaturated ketones and activated olefins to produce highly functionalized cyclopentanes. This study revealed that unsaturated ketones were very effective at delocalizing the radical away from the tin ketyl which was formed when tributyltin radical added to the ketone. Unactivated alkenes did not undergo cyclization reactions, but instead underwent simple reduction of the unsaturated ketone. Activated alkenes that were tethered to this O-stannyl allylic radical participated in 5-hexenyl radical cyclizations to form cyclopentanes. Excellent anti-stereoselectivities were achieved (>50:1) when these cyclizations were conducted at low concentrations. The effect of concentration on the product ratios seemed to indicate that these cyclizations were an artifact of a reversible cyclization. These observations are in direct contrast to how an α,β-unsaturated ketone is normally viewed in free radical reactions. These studies show that for the first time that unsaturated ketones can serve as radical precursors in a reagent-based study.

This study also demonstrated that the anionic character of the tin ketyl can be trapped by electrophilic reagents. All of the activating groups on the alkenes reacted with the tin enolate in a second cyclization when both appendages were syn to the newly formed cyclopentane. When the appendages were anti to one another, ring strain would not allow them to react, but electrophilic reagents such as bromine and deuterium oxide were able to trap this anti intermediate. This is important because the sequencing of one- and two-electron reactions is rapidly emerging as an important synthetic tool.[87]

Chapter 2 also demonstrated that the intermolecular coupling of α,β-unsaturated ketones was possible. Although this methodology appeared very limited to substrates bearing phenyl or radical-stabilizing groups, the implementation of either syringe pump or hydrideless tin techniques to these reactions may expand the applicability of this reaction.

The third chapter examined the reactions of epoxy ketones. The original goal of synthesizing tetrahydrofurans by this methodology was never realized. The well-known ability of alkoxy radicals to abstract hydrogen from adjacent carbons was not the problem with this project. It was a little-known 1,5-tin transfer mechanism which kept the alkoxy radical from cyclizing onto adjacent olefins. However, the carbon-centered radical, which was produced by this tin transfer, underwent cyclization reactions with tethered olefins. The product of these cyclizations were highly substituted cyclopentanols.

Collectively, these studies show that carbonyls can be effectively utilized as free-radical precursors. Hopefully this work will begin to establish the use of ketones and aldehydes in that regard. The cyclizations of unsaturated ketones and epoxy carbonyls have shown that these systems can offer excellent stereoselectivities along with a greater degree of manipulatable functionality. These cyclizations are well suited towards the synthesis of complex, and highly functionalized cyclopentanes, similar to those found in natural products. They will hopefully become a new and effective weapon in the arsenal of synthetic chemists.

CHAPTER 5

EXPERIMENTAL SECTION

General

Melting points were determined on a Thomas-Hoover
capillary melting point apparatus and are uncorrected.
Infrared spectra were recorded on a Perkin-Elmer 283B FTIR
spectrophotometer and are reported in wave numbers (cm^{-1}). ^1H
Nuclear magnetic resonance (NMR) spectra were recorded on a
Varian VXR-300 (300 MHz) spectrometer, and General Electric
QE-300 (300 MHz). ^{13}C NMR spectra were recorded at 75 MHz on
the above mentioned spectrophotometers. Chemical shifts are
reported in ppm downfield relative to tetramethylsilane
($(CH_3)_4Si$) as an internal standard in $CDCl_3$. Mass spectra and
exact mass measurements were performed on Finnigan MAT95Q,
Finnigan 4515, or Finnigan ITD mass spectrometers. Elemental
analysis was performed by Atlantic Microlab, Inc., Norcross,
GA 30091.

All reactions were run under an inert atmosphere of
argon using flame or heat dried apparatus. All reactions
were monitored by thin layer chromatography (TLC) and judged
complete when starting material was no longer visible in the
reaction mixture. All yields reported refer to isolated
material judged to be homogeneous by thin layer

chromatography and NMR spectroscopy. Temperatures above and below ambient temperature refer to bath temperatures unless otherwise stated. Solvents and anhydrous liquid reagents were dried according to established procedures by distillation under nitrogen from an appropriate drying agent: ether, benzene, and THF from benzophenone ketyl; CH_2Cl_2 from CaH_2. Other solvents were used "as received" from the manufacturer.

Analytical TLC was performed using Kieselgel 60 F-254 precoated silica gel plates (0.25 mm) using phosphomolybdic acid in ethanol as an indicator. Column chromatography was performed using Kieselgel silica gel 60 (230-400 mesh) by standard flash[89] and suction chromatographic techniques.

All GC experiments were performed on a Varian 3500 capillary gas chromatogarph using a J & W fused silica capillary column (DB5-30W; film thickness 0.25 μ).

Experimental Procedures and Results

(3E,8Z)-Tetradecadien-2-one (101)

To a previously dried 25 ml round bottom flask (RBF) was added methyl ketone Wittig reagent (1.70 g, 5.31 mmol) along with a magnetic stirrer. 5-Undecenal[90] (0.40 g, 2.38mmol) was weighed into a separate flask and transferred with 2.6 ml of $CHCl_3$. Reaction was quenched with water and extracted with Et_2O. The Et_2O layer was washed with sat. brine soln., dried over Na_2SO_4, and concentrated *in vacuo*. The concentrate was purified by flash chromatography on a silica gel column to yield a clear oil (0.40 g, 81.0%): R_f 0.75 (70% Et_2O/hexane); 300 MHz ^{1}H NMR ($CDCl_3$) δ 6.81 (1H, m), 6.08 (1H, d, J=16 Hz), 5.38 (2H, m), 2.26 (3H, s), 2.24 (2H, m), 2.03 (4H, m), 1.54 (2H, m), 1.30 (6H, m), 0.89 (3H, t, J=7 Hz); 75 MHz ^{13}C NMR ($CDCl_3$) δ 198.58, 148.23, 131.45, 131.01, 128.61, 32.58, 32.01, 31.56, 29.41, 28.18, 27.29, 26.29, 22.60, 14.08; IR (neat) 3006, 1700, 1628, 1459, 1253, 977 cm^{-1}; MS (CI), m/e (relative intensity) 209(m^++1, 9), 150(20), 137(21), 97(35), 95(28), 84(25), 81(34), 69(28), 67(29), 43(100); HR MS (CI) 209.1910 (calc. for $C_{14}H_{25}O$: 209.1905).

7-Oxo-5-octenal (104)[91]

Glutaric dialdehyde (100 ml of a 50% aq. solution, 552.3 mmol) was placed into a 300 ml RBF along with a magnetic stirrer. The methyl ketone Wittig reagent (25.5 g, 80.1 mmol) was dissolved in CH_2Cl_2 (100 ml) and placed into a large addition funnel. CH_2Cl_2 (50 ml) was added to the dialdehyde in the reaction flask and the ylide solution was slowly added to the reaction mixture. The addition of the ylide took about 1.5 hours, and the funnel was rinsed with 10 ml of CH_2Cl_2. The reaction was allowed to proceed overnight and was extracted with H_2O (2 x 100 ml). The organic layer was dried with Na_2SO_4 and concentrated *in vacuo*. Column chromatography of the residue produced a colorless oil (8.2784 g, 59.0 mmol, 74% yield): R_f 0.40 (70% Et_2O/hexane); 300 MHz ^{1}H NMR ($CDCl_3$)

δ 9.79 (1H, s), 6.77 (1H, dt, J=16.0, 6.9 Hz), 6.09 (1H, d, J=16.0 Hz), 2.52 (2H, t, J=6.9 Hz), 2.28 (5H, m), 1.84 (2H, m); 75 MHz ^{13}C NMR (CDCl$_3$) δ 201.51, 198.31, 146.65, 131.87, 42.98, 31.56, 26.95, 20.38.

General Procedure for Preparation of Activated Olefins

To a previously dried flask was added the appropriate stabilized ylide (7 mmol) and a magnetic stir bar. CHCl$_3$ (3.5 ml) was then added and the reaction was stirred. Once the ylide was completely dissolved, 7-oxo-5-octenal 104 (3.5 mmol) was added. Reaction was followed by TLC and starting material was usually consumed in 1-2 days. The reaction mixture was concentrated *in vacuo*, and this residue was flash chromatographed to yield the activated olefin.

9-Oxo-2E,7E-decadienenitrile (105)

Yield (47%); R_f 0.55 (90% Et$_2$O/hexane); 300 MHz ^{1}H NMR (CDCl$_3$) δ 6.75 (2H, m), 6.09 (1H, dt, J=16.0, 1.4 Hz), 5.38 (1H, dt, J=16.0, 1.4), 2.27 (7H, m), 1.68 (2H, m); 75 MHz ^{13}C NMR (CDCl$_3$) δ 198.24, 154.81, 146.36, 131.93, 117.28, 100.55, 32.60, 31.51, 27.06, 26.08; IR (neat) 2933, 2863, 2222 (s), 1731, 1696, 1673, 1631, 1363, 1256, 976 cm^{-1}; MS (EI), m/e (relative intensity) 164(M$^+$+1, self CI,18), 148(35), 55(54), 43(100), 41(33), 39(37); HR MS (EI) 163.09857 (calc. for C$_{10}$H$_{13}$NO: 163.09971).

3E,8E-Undecadiene-2,10-dione (106)

Yield (52%); R_f 0.55 (50% EtOAc/hexane); 300 MHz ^{1}H NMR (CDCl$_3$) δ 6.78 (2H, dt, J=16.0, 7.5 Hz), 6.04 (2H, d, J=16.0 Hz), 2.37 (10H, m), 1.67 (2H, m); 75 MHz ^{13}C NMR (CDCl$_3$) δ 198.27, 146.77, 131.78, 31.73, 27.02, 26.51; IR (neat) 2932, 1697, 1673, 1626, 1428, 1363, 1255, 1185, 977, 607 cm^{-1}; MS (EI), m/e (relative intensity) 180(M$^+$, 1), 137(43), 81(23), 43(100); HR MS (EI) 180.1151 (calc. for C$_{11}$H$_{16}$O$_2$: 180.11503).

Methyl-9-oxo-2E,7E-decadienoate (107)

Yield (49%); R_f 0.40 (70% Et$_2$O/hexane); 300 MHz ^{1}H NMR (CDCl$_3$) δ 6.95 (1H, dt, J=16.0, 6.9 Hz), 6.78 (1H, dt, J=16.0, 6.9 Hz), 6.08 (1H, dt, J=16.0, 1.5 Hz), 5.84 (1H, dt, J=16.0, 1.5 Hz), 3.73 (3H,s), 2.24 (7H, m), 1.67 (2H, m); 75 MHz ^{13}C NMR (CDCl$_3$) δ 198.29, 166.78, 148.13, 146.90, 131.63, 121.53, 51.34, 31.57, 31.38, 26.87, 26.28; IR (neat) 1723, 1697, 1674, 1627, 1436, 1362, 1272, 1256, 1202, 977 cm^{-1}; MS (EI), m/e (relative intensity) 197(M$^+$+1, self CI, 82), 196 (M$^+$, 2), 165(80), 164(31), 137(75), 136(55), 121(42), 93(61), 81(89), 79(28), 68(32), 55(51), 53(52), 43(100), 41(41), 39(56); HR MS (EI) 196.10913 (calc. for C$_{11}$H$_{16}$O$_3$: 196.10995).

General Procedure for Radical Cyclization of Olefins

A round bottom flask and a condenser were flame dried and allowed to cool under a argon atmosphere. After the flask had cooled, the activated olefin (0.5 mmol) was weighed into the flask. Then AIBN (0.05 mmol), followed by TBTH (1.5 mmol) and freshly distilled benzene (5 ml) were syringed into the flask. This solution was carefully degassed for 15 min with argon. Then the reaction was warmed to 85°C until the reaction was complete by TLC. When no starting material was observed by TLC, the reaction mixture was concentrated *in vacuo*, and the mixture was separated by flash chromatography to give the cis and trans cyclized products.

8Z-Tetradecen-2-one (102)

Yield (90%); R_f 0.77 (70% Et$_2$O/hexane); 300 MHz ^{1}H NMR (CDCl$_3$) δ 5.38(2H, m), 2.42 (2H, t, J=8 Hz), 2.14 (3H, s), 1.97 (4H, m), 1.58 (2H, m), 1.30 (10H, m), 0.89 (3H, t, J=7 Hz); 75 MHz ^{13}C NMR (CDCl$_3$) δ 209.0, 130.6, 129.9, 43.7, 32.3, 31.3, 29.4, 29.3, 28.8, 27.29, 26.9, 23.6, 22.5, 14.0; IR (neat) 2926, 2855, 1719, 1462, 1358, 1286, 1161, 1075, 968, 723 cm^{-1}; MS (CI), m/e (relative intensity) 211(m$^+$+1, 92.7), 193(32), 125(22), 85(41), 43(100); Anal. C$_{14}$H$_{26}$O: 79.82% C, 12.40% H (calc. 79.94% C, 12.46% H).

trans-(2-(2-Oxopropyl)cyclopentyl)ethanenitrile (108)

Yield (71%); R_f 0.60 (90% Et$_2$O/hexane); 300 MHz ^{1}H NMR (CDCl$_3$) δ 2.7-2.3 (4H, m), 2.14 (3H, s), 1.98 (3H, m), 1.77 (1H, m), 1.55 (2H, m), 1.44 (1H, m), 1.13 (1H, m); 75 MHz ^{13}C NMR (CDCl$_3$) δ 207.89, 119.15, 48.75 ,41.87, 40.28, 32.66, 32.06, 30.23, 23.55, 21.96; IR (neat) 2954, 2872, 2244 (s), 1713, 1452, 1424, 1359, 1294, 1227, 1174 cm^{-1}; MS (EI), m/e (relative intensity) 166(M^++1, self CI, 20), 81(24), 58(36), 43(100), 34(48); HR MS (EI) 165.11593 (calc. for $C_{10}H_{15}NO$: 165.11536).

cis-2-Acetyl-3-iminobicyclo[3.3.0]octane (109)

Yield (24%); R_f 0.40 (90% Et_2O/hexane); 300 MHz ^1H NMR ($CDCl_3$) δ 3.37 (1H, m), 2.70 (2H, m), 2.23 (1H, m), 2.10 (3H, s), 1.88 (1H, m), 1.74 (1H, m), 1.53 (4H, m), 1.36 (1H, m), 1.24 (1H, br s); 75 MHz ^{13}C NMR ($CDCl_3$) δ 196.19, 162.76, 110.90, 48.31, 41.29, 38.22, 34.87, 34.83, 27.95, 25.81; IR (KBr) 3355, 2929, 2856, 1627, 1498, 1340, 1316, 1284, 1270, 926 cm^{-1}; MS (EI), m/e (relative intensity) 165(M^+, 29), 136(100), 43(30), 34(52); HR MS (EI) 165.11526 (calc. for $C_{10}H_{15}NO$: 165.11536).

trans-1-(2-(2-Oxopropyl)cyclopentyl)-2-propanone (110)

Yield (73%); R_f 0.70 (50% EtOAc/hexane); 300 MHz ^1H NMR ($CDCl_3$) δ 2.60 (2H, dd, J=18, 5 Hz), 2.36 (2H, dd, J=18, 9 Hz), 2.14 (6H, s), 1.89 (4H, m), 1.58 (2H, m), 1.25 (2H, m); 75 MHz ^{13}C NMR ($CDCl_3$) δ 208.68, 48.97, 40.93, 32.35, 30.26, 23.50; IR (neat) 3602, 2949, 2871, 1713, 1407, 1359, 1274, 1230, 1176, 1154 cm^{-1}; MS (EI), m/e (relative intensity) 183(M^++1, self CI,100), 125(25), 124(45), 81(32), 43(95); HR MS (EI, self CI) 183.13866 (calc. for $C_{11}H_{19}O_2$: 183.13851).

cis-2(R)-Acetyl-3(S)-hydroxy-3-methylbicyclo[3.3.0]octane (111)

R_f 0.65 (50% EtOAc/hexane); 300 MHz ^{1}H NMR (CDCl$_3$) δ 4.09 (1H, br s), 2.86-2.67 (1H, d, J=9.3 Hz), 2.23 (3H, s), 2.00 (1H, dd, J=7.8, 13.2 Hz), 1.80-1.55 (6H, m), 1.31 (3H, s), 1.13 (1H, dd, 9.3, 13.2 Hz); 75 MHz ^{13}C NMR (CDCl$_3$) δ 214.64, 82.81, 66.34, 48.28, 47.68, 42.04, 33.26, 32.84, 32.15, 26.21, 25.63; IR (neat) 3474 (br), 2943, 2862, 1689, 1453, 1372, 1237, 1176, 1138, 934 cm^{-1}; MS (EI), m/e (relative intensity) 182(m$^+$, 0.1), 167(1), 124(71), 81(10), 66(20), 43(100); HR MS (CI) 183.1381 (calc. for C$_{11}$H$_{19}$O$_2$: 183.1385).

trans-Methyl(2-(2-oxopropyl)cyclopentyl)ethanoate (112)

Yield (58%); R_f 0.45 (70% Et$_2$O/hexane); 300 MHz ^{1}H NMR (CDCl$_3$) δ 3.67 (3H, s), 2.65-2.17 (4H, m), 2.14 (3H, s), 1.89 (4H, m), 1.59 (2H, m), 1.23 (2H, m); 75 MHz ^{13}C NMR (CDCl$_3$) δ 208.66, 173.66, 51.46, 48.91, 42.01, 40.76, 39.04, 32.33, 32.10, 30.27, 23.44; IR (neat) 2952, 2872, 1738, 1716, 1436, 1359, 1258, 1194, 1176, 1155 cm^{-1}; MS (CI), m/e (relative intensity) 199(M$^+$+1, 63), 167(92), 141(27), 81(60), 67(37), 43(100); HR MS (EI) 167.10746 (calc. for C$_{11}$H$_{16}$O$_3$ -OCH$_3$: 167.10721).

cis-2-Acetyl-3-oxobicyclo[3.3.0]octane (113)

Yield (27%); R_f 0.65 (70% Et$_2$O/hexane); 300 MHz ^{1}H NMR (CDCl$_3$) δ 13.8 (1H, br s), 3.26 (1H, m), 2.70 (2H, m), 2.06 (3H, s), 1.92 (3H, m), 1.58 (2H, m), 1.45 (2H, m); 75 MHz ^{13}C

NMR (CDCl$_3$) δ 202, 180.39, 115.27, 43.43, 42.93, 37.16, 34.80, 34.39, 26.04, 21.44; IR (neat) 2948, 2864, 1710, 1652, 1448, 1389, 1286, 1235, 937, 893 cm^{-1}; MS (EI), m/e (relative intensity) 166(M$^+$, 76), 137(100), 124(30), 95(48), 43(98), 41(31), 39(32); HR MS (EI) 166.09905 (calc. for C$_{10}$H$_{14}$O$_2$ 166.09938).

General Procedure for Enolate Trapping Studies

A round bottom flask and a condenser were flame dried and allowed to cool under a argon atmosphere. After the flask had cooled, activated olefin 107 (0..250 mmol) was weighed into the flask. Then AIBN (0.025 mmol), followed by TBTH (0.265 mmol) and freshly distilled benzene (25 ml) were syringed into the flask. This solution was carefully degassed for 15 min with argon. The reaction was warmed to 85°C for one hour, then either D$_2$O (XS) or Br$_2$ in CCl$_4$ (.750mmol) was added. When no starting material was observed on TLC, the reaction mixture was concentrated *in vacuo*, and the mixture was separated by flash chromatography to give mostly trans cyclized products.

Methyl(2-(1-bromo-2-oxopropyl)cyclopentyl)ethanoate (126)

Yield (86%); R_f 0.68 (90% Et$_2$O/hexane); 300 MHz ^{1}H NMR (CDCl$_3$) δ 4.45 (1H, d, J=7.5 Hz), 3.68 (3H, s), 2.47 (1H, dd, J=7.5, 15 Hz), 2.38 (3H, s), 2.27 (1H, dd, J=9, 15 Hz), 2.3-2.1 (2H, m), 1.90 (2H, m), 1.63 (2H, m), 1.45-1.2 (2H, m); 75 MHz ^{13}C NMR (CDCl$_3$) δ 201.96, 173.09, 60.27, 51.48, 46.41,

40.02, 38.98, 32.65, 30.16, 27.48, 23.81; IR (neat) 2953, 2871, 1736, 1436, 1358, 1257, 1198, 1166, 1086, 1014 cm^{-1}; MS (EI), m/e (relative intensity) 278(m$^+$, ^{81}Br,-), 276(m$^+$, ^{79}Br,-), 165(11), 141(16), 123(15), 95(17), 81(18), 43(100); HR MS (CI) 279.0417 (calc. for $C_{11}H_{18}{}^{81}BrO_3$: 279.0419).

Methyl(2-(1-deutero-2-oxopropyl)cyclopentyl)ethanoate (127)

Yield (87%), 85% Deuterium incorporation; R_f 0.60 (90% Et$_2$O/hexane); 300 MHz ^{1}H NMR (CDCl$_3$) δ 3.67 (3H, s), 2.62 (0.6H, dd, J=6, 18 Hz), 2.46 (1H, dd, J=6, 15 Hz), 2.35 (0.5H, dd, J=9, 18 Hz), 2.21 (1H, dd, J=9, 15 Hz), 2.14 (3H, s), 1.89 (4H, m), 1.59 (2H, m), 1.23 (2H, m); MS (CI), m/e (relative intensity) 200(m$^+$, 25), 199(24), 168(100), 167(99), 149(3), 140(12), 139(9), 121(4).

Hydrodimerization of Chalcone

Chalcone 132 (0.2057 g, 0.9877 mmol) was weighed into a previously dried 5 ml RBF. Freshly distilled Benzene (1 ml), a magnetic stir bar, AIBN (7.0 mg, 0.04 mmol), and TBTH (0.15 ml, 0.5577 mmol) were added and the reaction was degassed for 15 min. After TLC showed that chalcone was consumed, the reaction was concentrated *in vacuo*, and the residue was separated by column chromatography. The simple reduced product, 1,3-diphenyl-1-propanone,[92] 133 was obtained as a colorless solid which melted at 69-70°C (lit. 70-72°C). The hydrodimerized and cyclized product,[72] 134 was obtained as a mixture of isomers which was a colorless solid which melted at 93-101°C (lit. 91-99°C).

4,8-Dimethyl-3,7-nonadiene-2-ol (179)[93]

Geranial (mixture of citral and neral) (2.0327 g, 13.353 mmol) was weighed into a 50 ml RBF, then THF (25 ml) was added. The temperature was lowered to 0°C and 1.5M MeLi (13.5 ml, 20.25 mmol) was syringed into the reaction mixture. After an hour, the reaction was quenched with a sat. NH_4Cl soln., and then extracted with Et_2O (3 x 25 ml). The ether layer was dried over Na_2SO_4 and concentrated to give a colorless oil (2:1418 g, 12.728 mmol, 95.3% yield) which was a mixture of cis and trans double bond (spectra reported for major peaks of mixture): 300 MHz ^{1}H NMR ($CDCl_3$) δ 5.22 (1H, d, J=9 Hz), 5.09 (1H, t, J=7 Hz), 4.57 (1H, m), 2.08 (2H, m), 2.00 (2H, t, J=7 Hz), 1.88 (1H, br s), 1.69 (6H, s), 1.60 (3H, s), 1.22 (3H, d, J=8 Hz); 75 MHz ^{13}C NMR ($CDCl_3$) δ 137.23, 130.23, 129.17, 123.90, 64.64, 39.39, 26.35, 25.58, 23.55, 17.60, 16.33.

7,11-Dimethyl-6,10-dodecadiene-5-ol (180)[94]

Geranial (mixture of citral and neral) (1.9955 g, 13.108 mmol) was weighed into a 50 ml RBF, then THF (25 ml) was added. The temperature was lowered to 0°C and 2.5M BuLi (7.9 ml, 19.75mmol) was syringed into the reaction mixture. After an hour, the reaction was quenched with a sat. NH_4Cl soln., and then extracted with Et_2O (3 x 25 ml). The ether layer was dried over Na_2SO_4 and concentrated to give a colorless oil (2.65418 g, 12.617 mmol, 96.3% yield) which was a mixture of cis and trans double bond (spectra reported for major peaks of mixture): 300 MHz ^{1}H NMR ($CDCl_3$) δ 5.22-5.04 (2H, m), 4.33 (1H, m), 2.08 (5H, m), 1.69 (6H, s), 1.61 (3H, s), 1.31 (6H, m). 0.90 (3H, t, J=7 Hz); 75 MHz ^{13}C NMR ($CDCl_3$) δ 138.12, 131.53, 128.16, 123.92, 68.56, 39.53, 37.41, 27.58, 26.33, 25.60, 22.68, 17.62, 16.51, 14.02.

4,8-Dimethyl-3,7-nonadiene-2-one (182)[93]

Allylic alcohol 179 (2.1418 g, 12.728 mmol) was weighed into a 100 ml RBF with PDC (9.6 g, 25 mmol), CH_2Cl_2 (25 ml), and crushed 4Å molecular sieves.[95] The next day the reaction was diluted with Et_2O (75 ml) and this soln. was allowed to stir for four hours. Suction chromatography was used to separate the bulk of the PDC, and sieves. The eluent wss concentrated and purified by column chromatography to give a colorless oil (1.5105 g, 9.085 mmol, 71.4% yield) which was a mixture of cis and trans double bond (spectra reported for major peaks of mixture): 300 MHz ^{1}H NMR ($CDCl_3$) δ 6.08 (1H, s), 5.09 (1H, m), 2.16 (10H, m), 1.71 (3H, s), 1.62 (3H, s); 75 MHz ^{13}C NMR ($CDCl_3$) δ 198.59, 158.15, 132.35, 123.54, 122.96, 41.11, 31.65, 26.08, 25.57, 19.18, 17.59.

7,11-Dimethyl-6,10-dodecadiene-5-one (183)[96]

Allylic alcohol 180 (2.6541 g, 12.617 mmol) was weighed into a 100 ml RBF with PDC (9.5 g, 25 mmol), CH_2Cl_2 (25 ml), and crushed 4Å molecular sieves[95]. The next day the reaction was diluted with Et_2O (75 ml) and this soln. was allowed to stir for four hours. Suction chromatography was used to separate the bulk of the PDC, and sieves. The eluent was concentrated and purified by column chromatography to give a colorless oil (1.0289 g, 4.939 mmol, 39.1% yield) which is a mixture of cis and trans double bond (spectra reported for major peaks of mixture): 300 MHz ^{1}H NMR ($CDCl_3$) δ 6.05 (1H, s), 5.08 (1H, t, J=7.2 Hz), 2.60 (2H, t, J=8.1 Hz), 2.42 (2H, m), 2.14 (5H, m), 1.69 (3H, s), 1.62 (3H, s), 1.56 (2H, m), 1.33 (2H, m), 0.91 (3H, t, J=7,2 Hz); 75 MHz ^{13}C NMR ($CDCl_3$) δ 201.43, 157.77, 132.34, 123.88, 123.20, 44.11, 41.15, 33.74, 26.10, 25.60, 22.37, 19.23, 17.62, 13.86; IR (neat) 2960, 2930, 2873, 1688, 1619, 1450, 1377, 1131, 1076, 1040 cm^{-1}; MS (EI), m/e (relative intensity) 208(m^+,7), 166(22), 165(21), 151(70), 123(46), 98(41), 85(31), 83(100), 82(17), 69(38), 57(18), 55(17); HR MS (EI) 208.1856 (calc. for $C_{14}H_{24}O$: 208.1827).

3,7-Dimethyl-2,3-epoxy-7-octenal (185)

Unsaturated aldehyde 178 (0.9895 g, 6.500 mmol), MeOH (12 ml), 30% H_2O_2 in water (1.4 ml, 13.7 mmol), and a sat. aq. soln. of K_2CO_3 (2.9 ml) were added to a 25 ml RBF. The next day the reaction was quenched with sat. $NaHCO_3$. This soln. was extracted with Et_2O (3 x 25 ml), then the ether layer was dried over Na_2SO_4 and concentrated *in vacuo*. Column chromatography of the residue produced a colorless oil (0.291

g, 1.73 mmol, 26.6% yield): R_f 0.62 (50% Et$_2$O/Hexane); 300 MHz ^{1}H NMR (CDCl$_3$) δ 9.43 (1H, d, J=6 Hz), 5.07 (1H, t, J=7 Hz), 3.14 (1H, d, J=6 Hz), 2.12 (2H, m), 1.9-1.5 (11H, m); 75 MHz ^{13}C NMR (CDCl$_3$) δ 198.65, 132.54, 122.58, 64.39, 63.41, 38.18, 33.28, 25.51, 21.98, 17.06; IR (neat) 2970, 2928, 2858, 1723, 1674, 1452, 1409, 1383, 1110, 800 cm^{-1}; MS (CI), m/e (relative intensity) 169(m$^+$+1,33), 151(72), 137(57), 135(25), 123(100), 111(17), 109(66), 95(16), 82(15), 81(17); HR MS (CI) 169.1228 (calc. for C$_{10}$H$_{17}$O$_2$: 169.1226).

4,8-Dimethyl-3,4-epoxy-7-nonene-2-one (186)

Unsaturated ketone 182 (1.5105 g, 9.0851 mmol), MeOH (16 ml), 30% H$_2$O$_2$ in water (2.0 ml, 19.6 mmol), and a sat. aq. soln. of K$_2$CO$_3$ (3.3 ml) were added to a 25 ml RBF. The next day the reaction was quenched with sat. NaHCO$_3$. The reaction mixture was extracted with Et$_2$O (3 x 25 ml), then the ether layer was dried over Na$_2$SO$_4$ and concentrated *in vacuo*. Column chromatography of the residue produced a colorless oil (1.3225 g, 7.256 mmol, 79.9% yield): R_f 0.54 (50% Et$_2$O/hexane); 300 MHz ^{1}H NMR (CDCl$_3$) δ 5.11 (1H, t, J=7 Hz), 3.43 (1H, s), 2.21 (3H, s), 2.13 (2H, m), 1.85-1.50 (8H, m), 1.25 (3H, s); 75 MHz ^{13}C NMR (CDCl$_3$) δ 204.05, 132.36, 122.84, 64.74, 63.11, 38.06, 32.19, 27.80, 25.58, 23.60, 16.00; IR (neat) 2970, 2922, 1723, 1452, 1404, 1382, 1356, 1241, 1187, 1079 cm^{-1}; MS (CI), m/e (relative intensity) 183(m$^+$+1,11), 181(100), 165(24), 163(17), 143(47), 139(40), 137(10),

125(36), 123(11), 121(28); HR MS (CI) 183.1361 (calc. for $C_{11}H_{19}O_2$: 183.1385).

7,11-Dimethyl-6,7-epoxy-10-dodecene-5-one (187)

Unsaturated ketone 183 (0.9697 g, 4.654 mmol), MeOH (8 ml), 30% H_2O_2 in water (1.0 ml, 9.8 mmol), and a sat. aq. soln. of K_2CO_3 (1.8 ml) were added to a 25 ml RBF. The next day the reaction was quenched with sat. $NaHCO_3$. The reaction mixture was extracted with Et_2O (3 x 25 ml), then the ether layer was dried over Na_2SO_4 and concentrated *in vacuo*. Column chromatography of the residue produced a colorless oil (1.0176 g, 4.535 mmol, 97.5% yield): R_f 0.70 (50% Et_2O/hexane); 300 MHz ^1H NMR ($CDCl_3$) δ 5.11 (1H, t, J=7 Hz), 3.43 (1H, s), 2.50 (2H, m), 2.02 (2H, m), 1.87-1.47 (10H, m), 1.35 (2H, m), 1.22 (3H, s), 0.91 (3H, t, J+7 Hz); 75 MHz ^{13}C NMR ($CDCl_3$) δ 206.38, 132.46, 122.93, 64.53, 63.25, 40.54, 38.16, 32.17, 25.65 ,25.28, 23.67, 22.30, 16.05, 13.74; IR (neat) 2961, 2931, 2873, 1722, 1454, 1407, 1382, 1249, 1133, 1068 cm^{-1}; MS (CI), m/e (relative intensity) 224(m^+,1.8), 223(20), 205(2), 157(2), 143(6), 141(5), 139(8), 137(3), 125(8), 121(7); HR MS (CI) 223.1657 (calc. for $C_{14}H_{23}O_2$: 223.1698).

General Procedure for Cyclizations of Epoxy Ketones

The appropriate α,β-epoxy ketone (1.0 mmol) was weighed into a 10 ml pear-shaped flask (PSF). Then TBTH (2.0 mmol), AIBN (0.1 mmol), and benzene (2.0 ml) were added to the same flask. This mixture was degassed for 20 min. After it was degassed, the temperature was raised to 85°C. After the starting material had been consumed, the reaction was concentrated and purified by column chromatography to yield the cyclized products.

(2-Hydroxy-5-isopropyl-2-methylcyclopentyl)ethanone

This compound was isolated as a mixture of two products in a 3:2 ratio (^{1}H NMR integration). The isomers were not fully separated by column chromatography. The combined yield for both compounds was 82.7%. Despite the chromatography problems some fractions did contain pure compound and the spectra for each are reported below.

Major Product (189): R_f 0.31 (50% Et$_2$O/hexane); 300 MHz ^{1}H NMR (CDCl$_3$) δ 2.88 (1H, s), 2.65 (1H, d, J=10 Hz), 2.38 (1H, m), 2.27 (3H, s), 1.94 (1H, m), 1.83-1.48 (4H, m), 1.34 (3H, s), 0.84 (6H, dd, J=7, 12 Hz); 75 MHz ^{13}C NMR (CDCl$_3$) δ 213.67, 81.22, 64.61, 49.60, 41.45, 33.02, 32.07, 27.78, 26.08, 21.51, 19.38; IR (neat) 3446, 2959, 2872, 1699, 1466, 1422, 1369, 1245, 1170, 1129 cm^{-1}; MS (CI), m/e (relative intensity) 185(m$^+$+1,100), 184(1), 168(5), 167(50), 149(3), 141(2), 123(9), 109(1); HR MS (CI) 185.1526 (calc. for C$_{11}$H$_{21}$O$_2$: 185.1541).

Minor Product (190): R_f 0.25 (50% Et$_2$O/hexane); 300 MHz ^{1}H NMR (CDCl$_3$) δ 2.80 (1H, d, J=9 Hz), 2.26 (3H, s), 2.24 (1H, m), 1.75 (4H, m), 1.50 (2H, m), 1.18 (3H, s), 0.84 (6H, dd, J=7.5, 17 Hz); 75 MHz ^{13}C NMR (CDCl$_3$) δ 210.72, 81.59, 66.76,

47.00, 42.49, 32.97, 32.26, 26.03, 24.61, 21.29, 20.10; IR (neat) 3462, 2960, 2872, 1702, 1466, 1368, 1242, 1167, 1127, 932 cm^{-1}; MS (CI), m/e (relative intensity) 185(m$^+$+1,61), 169(2), 168(11), 167(100), 165(1), 149(8), 141(2), 124(1), 123(16), 109(3); HR MS (CI) 185.1545 (calc. for C$_{11}$H$_{21}$O$_2$: 185.1541).

1-(2-Hydroxy-5-isopropyl-2-methylcyclopentyl)-1-pentanone

This compound was isolated as a mixture of two products in a 1:1 ratio (by isolated weights). The combined yield for both compounds was 81.5%. The spectra for each are reported below.

Major Product (191): R_f 0.48 (50% Et$_2$O/hexane); 300 MHz ^{1}H NMR (CDCl$_3$) δ 3.10 (1H, s), 2.63 (1H, d, J=11 Hz), 2.54 (2H, t, J=7 Hz), 2.49 (1H, m), 2.00-1.50 (7H, m), 1.35 (2H, m), 1.31 (3H, s), 0.92 (3H, t, J=7 Hz), 0.85 (6H, dd, J=7, 13 Hz); 75 MHz ^{13}C NMR (CDCl$_3$) δ 215.94, 81.26, 63.83, 49.81, 45.86, 41.46, 31.74, 27.78, 25.80, 25.23, 22.28, 21.60, 19.14, 13.86; IR (neat) 3481, 2959, 2872, 1692, 1466, 1370, 1129, 1073, 942 cm^{-1}; MS (CI), m/e (relative intensity) 227(m$^+$+1,100), 219(6), 210(9), 209(65), 191(8), 183(4), 149(7), 127(11), 124(4), 123(55); HR MS (CI) 227.2015 (calc. for C$_{14}$H$_{27}$O$_2$: 227.2011).

Minor Product (192): R_f 0.40 (50% Et$_2$O/hexane); 300 MHz ^{1}H NMR (CDCl$_3$) δ 2.78 (1H, d, J=9 Hz), 2.60-2.41 (2H, m), 2.23 (1H, m), 1.92-1.42 (8H, m), 1.32 (2H, m), 1.15 (3H, m), 0.93 (3H, t, J=8 Hz), 0.83 (6H, dd, J=7, 18 Hz); 75 MHz ^{13}C NMR

(CDCl$_3$) δ 212.62, 81.75, 66.07, 47.18, 44.83, 42.50, 32.82, 26.01, 25.41, 24.80, 22.30, 21.33, 20.02, 13.89; IR (neat) 3474, 2958, 2872, 1694, 1466, 1377, 1307, 1244, 1109, 1068 cm^{-1}; MS (CI), m/e (relative intensity) 227(m$^+$+1,11), 226(2), 214(1), 211(1), 210(6), 209(53), 208(1); HR MS (CI) 227.1999 (calc. for C$_{14}$H$_{27}$O$_2$: 227.2011).

2-Hydroxy-5-isopropyl-2-methylcyclopentanecarboxaldehyde

This compound was isolated as a mixture of two products in a 3:1 ratio (^{1}H NMR integration and Capillary GC). The isomers were not fully separated by column chromatography, and the combined yield for both compounds was 81%. Despite the chromatography problems some fractions did contain pure compound and the spectra for each are reported below.

Major Product (198): R_f 0.31 (50% Et$_2$O/hexane); 300 MHz ^{1}H NMR (CDCl$_3$) δ 9.80 (1H, s), 2.46 (1H, m), 2.30 (1H, d, J=9 Hz), 2.13 (2H, m), 1.8-1.4 (4H, m), 1.42 (3H, s), 0.88 (6H, dd, J=7, 13 Hz); 75 MHz ^{13}C NMR (CDCl$_3$) δ 206.01, 82.78, 64.42, 46.78, 41.67, 32.26, 27.85, 26.83, 21.18, 19.99; IR (neat) 3437, 2960, 2872, 2730, 1715, 1466, 1378, 1268, 1136, 1074 cm^{-1}; MS (CI), m/e (relative intensity) 171(m$^+$+1,7), 169(4), 155(8), 154(5), 153(69), 137(12), 135(23), 127(12), 114(4), 100(11); HR MS (CI) 173.1354 (calc. for C$_{10}$H$_{19}$O$_2$: 171.1385).

Minor Product (199): R_f 0.25 (50% Et$_2$O/hexane); 300 MHz ^{1}H NMR (CDCl$_3$) δ 9.70 (1H, d, J=3 Hz), 2.54 (1H, dd, J=3, 7 Hz), 2.18 (1H, m), 1.9-1.5 (5H, m), 1.32 (3H, s), 1.25 (1H, s), 0.88 (6H, dd, J=7, 12 Hz); 75 MHz ^{13}C NMR (CDCl$_3$) δ 203.93, 66.62, 45.97, 42.10, 32.92, 29.70, 27.08, 25.04, 21.31, 20.33; IR (neat) 3430, 2957, 2872, 2729, 1716, 1463, 1376, 1075, 926 cm^{-1}; MS (CI), m/e (relative intensity) 171(m^{+}+1,40), 155(17), 153(100), 135(22); HR MS (CI) 171.1351 (calc. for C$_{10}$H$_{19}$O$_2$: 171.1385).

7-Octene-1,6-diol (212)

ε-Caprolactone (5.21 g, 45.6 mmol) and CH$_2$Cl$_2$ (92 ml) were added to a 250 ml RBF equipped with a magnetic stir bar. This solution was chilled to -78°C with a dry ice/acetone slush bath. DIBAL (50 ml of a 1M soln., 50.0 mmol) was cautiously added over a 20 min. into the flask by placing the syringe needle tip onto the upper neck of flask so that DIBAL did not drip directly, but rather ran down the flask and was cooled before it gets to solution. After 1 hr. the reaction was quenched with MeOH (3 ml) at -78°C. The reaction mixture was then poured into a freshly prepared aq. sat. Rochelle's salt soln. (300 ml) with rapid stirring, and this was allowed to stir overnight. The reaction mixture was extracted with EtOAc (4 x 150 ml), then the organic layer was dried over Na$_2$SO$_4$ and concentrated to yield crude 6-hydroxyhexanal. The crude product polymerized readily so it was used without further purification.

The crude 6-hydroxyhexanal was placed in a 500 ml RBF along with freshly distilled THF (150 ml) and a magnetic stirrer. This mixture was chilled to 0°C with an ice bath.

A 1M solution of vinyl magnesium bromide in THF (150 ml, 150.0 mmol) was added to a large addition funnel by cannula. The Grignard reagent was slowly dripped into the reaction mixture over approximately a 1 hour period. The reaction was allowed to slowly warm to room temperature overnight, and it was cautiously quenched with a sat. NH_4Cl soln. the next day. The reaction mixture was extracted with EtOAc (2 x 100 ml), and the combined organic layers were dried over Na_2SO_4. The concentrate was purified by flash chromatography on a silica gel column to yield a clear oil.(3.35 g, 51%, from ε-caprolactone): R_f 0.39 (80% EtOAc/hexane); 300 MHz ^1H NMR ($CDCl_3$) δ 5.87 (1H, m), 5.22 (1H, d, J=17 Hz), 5.09 (1H, d, J=17 Hz), 4.09 (1H, m), 3.63 (2H, t, J=7 Hz), 2.10 (2H, m), 1.57 (4H, m), 1.39 (4H, m); 75 MHz ^{13}C NMR ($CDCl_3$) δ 141.5, 114.2, 72.9, 62.7, 37.0, 32.5, 26.1, 25.2; IR (neat) 3849, 3812, 3356, 2934, 2860, 1644, 1426, 1055, 994, 922 cm^{-1}; MS (CI), m/e (relative intensity) 145(m^++1, 6), 143(7), 127(40), 125(4), 110(8), 109(100), 99(4), 85(4), 71(4), 67(5); HR MS (CI) 145.1225 (calc. for $C_{11}H_{17}O_2$: 145.1229).

7,8-Epoxyoctane-1,6-diol (213)

Diol 212 (0.9865 g, 6.840 mmol) was weighed into a 100 ml RBF and CH_2Cl_2 (20 ml) was added. 50% mCBPA (2.66 g, 7.71 mmol) was weighed into a separate RBF and transferred with CH_2Cl_2 (15 ml) into the 100 ml RBF. The next day the cloudy solution was suction chromatographed to yield a crude product which was further purified by column chromatography to give a colorless oil (0.863 g, 5.39 mmol, 79% yield): R_f 0.22 (80% EtOAc/hexane); 300 MHz ^{1}H NMR (CDCl$_3$) δ 3.64 (2H, t, J=7 Hz), 3.42 (1H, m), 2.98 (1H, m), 2.82 (1H, m), 2.71 (1H, m), 2.53 (2H, br s), 1.7-1.3 (8H, m); 75 MHz ^{13}C NMR (CDCl$_3$) δ 71.60, 62.47, 55.56, 45.11, 34.07, 32.42, 25.64, 25.01; IR (neat) 3387, 2935, 2861, 1656, 1462, 4256, 1054, 912, 854, 802 cm^{-1}; MS (CI), m/e (relative intensity) 161(m$^+$+1,14), 144(8), 143(100), 125(49), 107(22), 99(6), 97(4), 95(11), 81(13), 71(3); HR MS (CI) 161.1183 (calc. for $C_8H_{17}O_3$: 161.1178).

7,8-Epoxy-6-oxooctanal (214)

The Dess-Martin Periodinane[87] was weighed into a 50 ml RBF while being blanketed with an argon atmosphere, then CH$_2$Cl$_2$ (13 ml) was added. Diol 213 (0.7564 g, 4.721 mmol) was weighed into a separate flask and transferred into the reaction with CH$_2$Cl$_2$ (6 ml). After addition of 213 the reaction rapidly boiled for about 30 sec. After 5 min, the starting material was consumed and the reaction mixture was diluted with 25 ml of Et$_2$O. This mixture was poured into 25 ml of Et$_2$O containing Na$_2$SO$_3$ (6.5 g). After the exothermic quench was complete the reaction mixture was diluted with 25 ml of Et$_2$O, then extracted with NaHCO$_3$ (1 x 25 ml), and H$_2$O (1 x 25 ml). The organic layer was dried over Na$_2$SO$_4$ and concentrated *in vacuo* to yield a colorless oil (0.402 g, 2.54 mmol, 54% yield): R_f 0.48 (80% EtOAc/hexane); 300 MHz ^{1}H NMR (CDCl$_3$) δ 9.77 (1H, s), 3.43 (1H, dd, J=3, 5 Hz), 3.11 (1H, dd, J=5, 6 Hz), 2.88 (1H, dd, J=5, 6 Hz), 2.55-2.25 (4H, m),

1.63 (4H, m); 75 MHz ^{13}C NMR (CDCl$_3$) δ 213.81, 201.93, 53.36,, 46.08, 43.54, 36.06, 22.39, 21.39; IR (neat) 2937, 1714, 1402, 1238, 871 cm^{-1}; MS (CI), m/e (relative intensity) 157(m$^+$+1,2), 156(2), 155(35), 139(1), 137(6), 129(5), 115(1), 113(2), 111(1), 109(1); HR MS (CI) 157.0832 (calc. for C$_8$H$_{13}$O$_3$: 157.0865).

Methyl-9,10-epoxy-8-oxo-2-decenoate (215)

Aldehyde 214 (0.402 g, 2.54 mmol) was weighed into a 25 ml RBF. Then CH$_2$Cl$_2$ (5.0 ml) and methyl ester Wittig reagent (1.5 g, 4.5 mmol) were added to the flask and the reaction ran overnight. The next day the reaction was concentrated and purified by column chromatography to yield a colorless oil (0.3640 g, 1.719 mmol, 68% yield): R_f 0.51 (50% EtOAc/hexane); 300 MHz ^{1}H NMR (CDCl$_3$) δ 6.94 (1H, dt, J=6.9, 15.6 Hz), 5.82 (1H, d, J=15.6 Hz), 3.72 (3H, s), 3.42 (1H, m), 3.00 (1H, m), 2.87 (1H, m), 2.38 (2H, m), 2.22 (2H, m), 1.61 (2H, m), 1.46 (2H, m); 75 MHz ^{13}C NMR (CDCl$_3$) δ 207.19,, 166.89,, 148.65, 121.21, 53.32, 51.30, 46.00, 36.00, 31.81, 27.37, 22.40; IR (neat) 2949, 2861, 1721, 1656, 1437, 1274, 1199, 1035, 983, 871 cm^{-1}; MS (CI), m/e (relative intensity) 213(m$^+$+1,14), 201(1), 195(12), 181(1), 177(1), 163(4), 141(9); HR MS (CI) 213.1112 (calc. for C$_{11}$H$_{17}$O$_4$: 213.1127).

Methyl-10-hydroxy-8-oxo-decenoate (216)

Epoxy ketone 215 (40.4 mg, 0.1903 mmol) was weighed into a 50 ml RBF. Then AIBN (8.2 mg, 0.05 mmol), TBTH (0.98 ml,

3.6 mmol), and freshly distilled benzene (23 ml) were added to the flask. This solution was degassed for 20 min, then it was warmed to 85°C. After three hours, the reaction mixture was cooled and concentrated *in vacuo*. This mixture was separated by column chromatography to yield a colorless oil (28.6 mg, 0.134 mmol, 70% yield: R_f 0.22 (50% EtOAc/hexane); 300 MHz ^{1}H NMR (CDCl$_3$) δ 6.94 (1H, dt, J=6.9, 15.6 Hz), 5.82 (1H, d, J=15.6 Hz), 3.85 (2H, t, J=5.1 Hz), 3.73 (3H, s), 2.66 (2H, t, 5.7 Hz), 2.52 (1H, br s), 2.47 (2H, t, J=7.2 Hz), 2.21 (2H, m), 1.61 (2H, m), 1.47 (2H, m); 75 MHz ^{13}C NMR (CDCl$_3$) δ 211.17, 167.05, 148.78, 121.32, 57.87, 51.42, 44.45, 42.96, 31.93, 27.51, 23.00; IR (neat) 3477, 2949, 1720, 1656, 1438, 1317, 1274, 1202, 1177, 1042 cm^{-1}; MS (CI), m/e (relative intensity) 215(m$^+$+1,100), 197(25), 179(13), 167(6), 165(17), 147(9), 141(7), 137(6), 135(3), 119(11); HR MS (CI) 215.1293 (calc. for C$_{11}$H$_{19}$O$_4$: 215.1283).

LIST OF REFERENCES

1. Giese, B. *Radicals in Organic Synthesis: Formation of Carbon--Carbon Bonds;* Pergamon: New York, 1986.

2. Motherwell, W.B.; Crich, D. *Free Radical Chain Reactions in Organic Synthesis;* Academic: New York, 1992.

3. Pine, S.H. *Organic Chemistry;* McGraw-Hill: New York, 1987.

4. Gomberg, M. *J. Am. Chem.Soc.* **1900**, *22*, 757.

5. Gomberg, M. *Chem. Ber.* **1900**, *33*, 3150.

6. Paneth, F.; Hofeditz, W. *Chem. Ber.* **1929**, *62*, 1335.

7. Hey, D.H.; Waters, W.A. *Chem. Rev.* **1937**, *21*, 169.

8. Kharasch, M.S.; Margolis, E.T.; Mayo, F.R. *J. Org. Chem.* **1937**, *2*, 393.

9. Mayo, F.R.; Lewis, F.M. *J. Am. Chem. Soc.* **1944**, *66*, 1594.

10. Mayo, F.R.; Lewis, F.M. *Discuss. Faraday Soc.* **1947**, *2*, 285.

11. Bunnett, J.F. *Acc. Chem. Res.* **1978**, *11*, 413.

12. Minisci, F. *Top. Curr. Chem.* **1976**, *62*, 1.

13. (a) Neumann, W.P. *Synthesis* **1987**, *8*, 665. (b) Thebtaranonth, C.; Thebtaranonth, Y. *Tetrahedron* **1990**, *46*, 1385. (c) Hart, D.J. *Science* **1984**, *223*, 883.

14. Curran, D.P. In *Advances in Free Radical Chemistry;* Tanner, D.D., Ed.; JAI: Greenwich, Connecticut, 1990; Chapter 3.

15. Curran, D.P.; Rakiewicz, D.M. *J. Am. Chem. Soc.* **1985**, *107*, 1448.

16. Curran, D.P.; Chen, M.H. *Tetrahedron Lett.* **1985**, *26*, 4991.

17. Curran, D.P.; Fevig, T. L.; Elliott, R. L. *J. Am. Chem. Soc.* **1988**, *110*, 5064.

18. Curran, D.P.; Jasperse, C.P. *J. Am. Chem. Soc.* **1990**, *112*, 5601.

19. Curran, D.P.; Kuo, S.C. *J. Am. Chem. Soc.* **1986**, *108*, 1106.

20. Menapace, L.W.; Kuiviala, H.G. *J. Am. Chem. Soc.* **1964**, *86*, 3047.

21. Pereyre, M.; Quintard, J.P.; Rahm, A. *Tin in Organic Synthesis*; Butterworths: London, 1987.

22. Curran, D.P. *Synthesis* **1988**, *6*, 417, 489.

23. (a) Curran, D.P. *Chem. Rev.* **1991**, *91*, 1237. (b) Giese, B. Angew. *Chem. Int. Ed. Engl.* **1985**, *24*, 553. (c) Ramaiah, M. *Tetrahedron* **1987**, *43*, 3541.

24. Keinan, E.; Peretz, M. *J. Org. Chem.* **1983**, *48*, 5302.

25. Weinshenker, N.M.; Crosby, G.A.; Wong, J.Y. *J. Org. Chem.* **1975**, *40*, 1966.

26. van der Kerk, G.J.M.; Noltes, J.G.; Luitjen, J.G.A. *J. Appl. Chem.* **1957**, *7*, 356.

27. Davies, A.G. In *Chemistry of Tin*; Harrison, P.G., Ed.; Chapman and Hall: New York, 1989; Chapter 9.

28. Gutierrez, C.G.; Summerhays, L.R. *J. Org. Chem.* **1984**, *49*, 5206.

29. Barton, D.H.R.; McCombie, S.W. *J. Chem. Soc. Perkin 1* **1975**, 1574.

30. Rosen, T.; Taschner, M.J.; Heathcock, C.H. *J. Org. Chem.* **1984**, *49*, 3994.

31. Fraser-Reid, B.; Tulshian, D.B. *Tetrahedron Lett.* **1980**, *21*, 4549.

32. Beale, M.H.; Gaskin, P.; Kirkwood, P.S.; MacMillan, J. *J. Chem. Soc., Perkin Trans.1* **1980**, 885.

33. Matlin, S.A.; Gandham, P.S. *J. Chem. Soc., Chem. Commun.* **1984**, 798.

34. Yang, T.X.; Four, P.; Guire, F.; Balavoine, G. *Nouv. J. Chim.* **1984**, *8*, 6111.

35. Tanner, D.D.; Gielen, M.; Potter, A. *J. Org. Chem.* **1985**, *50*, 2149.

36. Beckwith, A.L.J.; Roberts, D.H. *J. Am. Chem. Soc.* **1986**, *108*, 5893.

37. Sugawara, T.; Otter, B.A.; Ueda, T. *Tettrahedron Lett.* **1988**, *29*, 75.

38. Giese, B.; Kretzschmar, G. *Chem. Ber.* **1983**, *116*, 3267.

39. Fleming, I. *Frontier Orbitals and Organic Chemical Reactions*; Wiley: London, 1976.

40. Beckwith, A.L.J. *Tetrahedron* **1981**, *37*, 3073.

41. (a) Beckwith, A.L.J.; Schiesser, C.H. *Tetrahedron* **1985**, *41*, 3925. (b) Beckwith, A.L.J.; Phillipou, G.; Serelis, A.K. *Tetrahedron Lett.* **1981**, *22*, 2811. Beckwith, A.L.J.; Meijs, G.F. *J. Chem. Soc. Perkins Trans. 2* **1979**, 1535.

42. Dewar, M.J.S.; Olivella, S. *J. Am. Chem. Soc.* **1978**, *100*, 5290.

43. Fujimoto, H.; Yamabee, S.; Minato, T.; Fukui, K. *J. Am. Chem. Soc.* **1972**, *94*, 9205.

44. Neumann, W.P. *Tetrahedron* **1989**, *45*, 951.

45. Stork, G.; Meisels, A.; Davies, J.E. *J. Am. Chem. Soc.* **1963**, *85*, 3419.

46. Schafer, H.J.; Seidel, W.; Knolle, J. *Chem. Ber.* **1977**, *110*, 3544.

47. Giese, B.; Gonzalez-Gomez, J.A.; Witzel, T. *Angew. Chem. Int. Eng. Ed.* **1984**, *23*, 69.

48. Giese, B.; Horler, H.; Leising, M. *Chem. Ber.* **1986**, *119*, 444.

49. Porter, N.A.; Swann, E.; Nally, J.; McPhail, A. *J. Am. Chem. Soc.* **1990**, *112*, 6740.

50. Giese, B.; Zehnder, M.; Roth, M.; Zietz, H.G. *J. Am. Chem. Soc.* **1990**, *112*, 6741.

51. Curran, D.P.; Shen, W.; Zhang, J.; Heffner, T. *J. Am. Chem. Soc.* **1990**, *112*, 6738.

52. Porter, N.A.; Scott, D.M.; McPhail, T. *Tetrahedron Lett.* **1990**, *31*, 1679.

53. Nonhebel, D.C.; Walton, J.C. *Free-Radical Chemistry*; Cambridge University: Cambridge, England, 1974.

54. Rawal, V.H.; Newton, R.C.; Krishnamurthy, V. *J. Org. Chem.* **1990**, *55*, 5181.

55. Motherwell, W.B.; Harling, J.D. *J. Chem. Soc. Chem. Commun.* **1988**, 1380.

56. Motherwell, W.B. *Aldrichimica Acta* **1992**, *25*, 71.

57. Kochi, J.K., Ed., *Free Radicals*; John Wiley and Sons: New York, 1973.

58. Fry, A.J. *Synthetic Organic Electrochemistry*; John Wiley and Sons: New York, 1989.

59. Quinkert, G.; Opitz, K.; Weirsdorff, W.W.; Weinlich, J. *Tetrahedron Lett.* **1963**, 1863.

60. Carey, F.A.; Sundberg, R.J. *Advanced Organic Chemistry*; Plenum: New York, 1984.

61. Shono, T.; Nishiguchi, I.; Ohmizu, H. *Chem. Lett.* **1966**, 1233.

62. Beckwith, A.L.J.; Moad, G.J. *J. Chem. Soc. Chem Commun.* **1974**, 472.

63. Enholm, E.J.; Prasad, G. *Tetrahedron Lett.* **1989**, *30*, 4939.

64. Enholm, E.J.; Burroff, J.A. *Tetrahedron Lett.* **1992**, *33*, 1835.

65. Enholm, E.J.; Kinter, K.S. *J. Am. Chem. Soc.* **1991**, *113*, 7784.

66. There are only a few examples of a sodium-promoted intermolecular coupling of the β-carbons of activated olefins. Most cases involve enones blocked on both ends (usually with t-butyl groups).

67. (a) Little, D.; Bazier, M. J. *J. Org. Chem.* **1986**, *51*, 4497. (b) Margaretha, P. *Helv. Chim. Acta.* **1982**, *65*, 1949. (c) *The Chemistry of Enones*; Patai, S.; Rappaport, Z., Eds.; John Wiley and Sons: New York, 1989.

68. Grimshaw, J.; Stevenson, G.R. *J. Am. Chem. Soc.* **1971**, *93*, 2432.

69. Pattenden, G. *Tetrahedron Lett.* **1987**, *28*, 1313.

70. Kigoshi, H.; Imamura, Y.; Niwa, H.; Yamada, K. *J. Am. Chem. Soc.* **1989**, *111*, 2302.

71. Tissot, P.; Surbeck, J.P.; Gulacar, F.O.; Margaretha, P. *Helv. Chim. Acta.* **1981**, *64*, 1570.

72. Fournier, F.; Berthelot, J.; Basselier, J.J. *Tetrahedron* **1985**, *41*, 5667.

73. Ingold, K.U. *Pure Appl. Chem.* **1984**, *56*, 1767.

74. Laird, E.R.; Jorgensen, W.L. *J. Org. Chem.* **1990**, *55*, 9.

75. Feldman, K.S.; Fisher, T.E. *Tetrahedron* **1989**, *45*, 2969.

76. (a) Hasegawa, E.; Ishiyama, K.; Horaguchi, T.; Shimizu, Y. *J. Chem. Soc. Chem. Commun.* **1990**, 550. (b) Hasegawa, E.; Ishiyama, K.; Kato, T.; Horaguchi, T.; Shimizu, Y. *J. Org. Chem.* **1992**, *57*, 5352.

77. (a) Stogryn, E.L.; Gianni, M.H. *Tetrahedron Lett.* **1970**, 3025. (b) Cook, M.; Hares, O.; Johns, A.; Murphy, J.A.; Patterson, C.W. *J. Chem. Soc., Chem. Commun.* **1986**, 1419. (c) Dickinson, J.M.; Murphy, J.A.; Patterson, C.W.; Woooster, N.F. *J. Chem. Soc., Perkin Trans. 1* **1990**, 1179.

78. (a) Brown, H.C.; Midland, M.M. *J. Am. Chem. Soc.* **1971**, *93*, 4078. (b) Barton, D.H.R.; Hay-Motherwell, R.S.; Motherwell, W.B. *J. Chem. Soc., Perkin Trans. 1* **1981**, 2363. (c) Bowman, W.R.; Marples, B.A.; Zaidi, N.A. *Tetrahedron Lett.* **1989**, *30*, 3343.

79. Wharton, P.S. *J. Org. Chem.* **1961**, *26*, 4781.

80. This addition method was not used in our research, but was included to show a different method for conducting these reactions.

81. Johns, A.; Murphy, J.A. *Tetrahedron Lett.* **1988**, *29*, 837.

82. Barton, D.H.R.; Beaton, I.M.; Geller, L.E.; Pechet, M.M. *J. Am. Chem. Soc.* **1960**, *82*, 2640.

83. Kim, S.; Lee, S.; Koh, J.S. *J. Am. Chem. Soc.* **1991**, *113*, 5106.

84. Davies, A.G.; Tse, M.W. *J. Organomet. Chem.* **1978**, *155*, 25.

85. Bowman, W.R.; Brown, D.S.; Burns, C.A.; Marples, B.A.; Zaidi, N.A. *Tetrahedron* **1992**, *48*, 6883.

86. (a) Beckwith, A.L.J.; Phillipou, G. *Aust. J. Chem.* **1976**, *29*, 123. (b) Beckwith, A.L.J.; Moad, G. *J. Chem. Soc., Perkin Trans. 2* **1980**, 1473. (c) Beckwith, A.L.J. *Tetrahedron* **1981**, *37*, 3073.

87. Dess, D.B.; Martin, J.C. *J. Am. Chem. Soc.* **1991**, *113*, 7277.

88. Curran, D.P.; Fevig, T.L.; Totleben, M. *Synlett* **1990**, 733.

89. Still, W.; Kahn, M.; Mitra, A. *J. Org. Chem.* **1978**, *43*, 2923.

90. Wilson, S.R.; Zucker, P.A.; Kim, C.W.; Villa, C.A. *Tetrahedron Lett.* **1985**, *26*, 1969.

91. Kovalev, B.G.; Dormindontova, N.P.; Shamshurin, A.A. *Zh. Org. Khim.* **1969**, *5*, 1775.

92. Hamrick, P.J. Jr.; Hauser, C.R. *J. Am. Chem. Soc.* **1959**, *81*, 493.

93. Masuyama, Y.; Takahashi, M.; Kurusu, Y. *Tetrahedron Lett.* **1984**, *25*, 4417.

94. CAS Registry # 91418-29-0.

95. Corey, E.J. *Tetrahedron Lett.* **1979**, *20*, 399.

96. CAS Registry # 91418-32-5.

BIOGRAPHICAL SKETCH

Kevin Kinter was born in York, Pennsylvania, the fifth son of Mernee and Jim Kinter. He moved to Bowie, Maryland, when he was only a few months old and finally settled down in Richmond, Virginia, after a few years. He grew up in the quiet suburbs of one the larger cities in Virginia. He spent a great deal of time wandering through the woods, lakes, and creeks of the surrounding community searching for snakes, lizards, turtles and many other undesirable pets. His love for God's creatures spawned dreams of one day becoming a vet.

He began his education at Pinchback Elementary School, which was a mere five minute walk through the woods. He was an unruly and inattentive child who found recess and nap time to be the most gratifying aspects of school. It was his sixth grade teacher, Mrs. Evans, who began to interest him in the education process. At Byrd Middle School he began to gain an appreciation for his math and science courses. This may have been due to his wrestling coach, Mr. Beard, being both a physical science teacher and an avid snake fan.

He went to high school at Douglas Freeman, and although he was a little shy, he led an active social life. He also enjoyed wrestling and was a four year varsity letter winner. The dissection of animals in his tenth grade biology class demonstrated that becoming a vet was not a good idea. The next year he took an honors chemistry course with Mr. Homer E. Alberti and immediately recognized that this was his future profession.

He went to the same undergraduate university, James Madison University in Harrisonburg, Virginia, that all four of his older brothers attended. In his freshman year he did a semester of research for Dr. Frank Palocsay. The next year he began doing research for Dr. Gary Crowther on the metal-halogen exchange reactions, which continued until he graduated in June of 1988.

Yearning to get away from the cold, he decided to attend the University of Florida, where he joined Dr. Eric Enholm's research group with a wide-eyed enthusiasm. However, he soon learned that graduate research was a little more challenging than his undergraduate projects. After many years of frustration and progress, he now finds himself on the verge of the next giant step. He will now travel to Duke University in Durham, North Carolina, where he will work for Dr. Ned Porter. In some ways he truly feels a long way from his old home in Richmond, but the excitement and anticipation of a new city and new chemistry remind him of the day he left home for college.

I certify that I have read this study and that in my opinion it conforms to acceptable standards of scholarly presentation and is fully adequate, in scope and quality, as a dissertation for the degree of Doctor of Philosophy.

Eric J. Enholm, Chairman
Assistant Professor of Chemistry

I certify that I have read this study and that in my opinion it conforms to acceptable standards of scholarly presentation and is fully adequate, in scope and quality, as a dissertation for the degree of Doctor of Philosophy.

Merle A. Battiste
Professor of Chemistry

I certify that I have read this study and that in my opinion it conforms to acceptable standards of scholarly presentation and is fully adequate, in scope and quality, as a dissertation for the degree of Doctor of Philosophy.

William M. Jones
Distinguished Service
Professor of Chemistry

I certify that I have read this study and that in my opinion it conforms to acceptable standards of scholarly presentation and is fully adequate, in scope and quality, as a dissertation for the degree of Doctor of Philosophy.

Randolph S. Duran
Assistant Professor of
Chemistry

I certify that I have read this study and that in my opinion it conforms to acceptable standards of scholarly presentation and is fully adequate, in scope and quality, as a dissertation for the degree of Doctor of Philosophy.

_____
Kenneth B. Sloan
Associate Professor of
Medicinal Chemistry

This dissertation was submitted to the Graduate Faculty of the Department of Chemistry in the College of Liberal Arts and Sciences and to the Graduate School and was accepted as partial fulfillment of the requirements for the degree of Doctor of Philosophy.

May, 1993

_____
Dean, Graduate School